Modern Antenna

The modern antenna book series mainly covers the related antenna theories and technologies proposed and studied in recent years to solve the bottleneck problems faced by antennas, including binary coded antenna optimization method, artificial surface plasmon antenna, complex mirror current equivalent principle and low profile antenna, generalized pattern product principle and generalized antenna array, cross dielectric transmission antenna, metamaterial antenna, as well as new antenna technology and development. This series not only presents the important progress of modern antenna technology from different aspects, but also describes new theoretical methods, which can be used in modern and future wireless communication, radar detection, internet of things, wireless sensor networks and other systems.

The purpose of the modern antenna book series is to introduce new antenna concepts, new antenna theories, new antenna technologies and methods in recent years to antenna researchers and engineers for their study and reference. Each book in this series is thematic. It gives a comprehensive overview of the research methods and applications of a certain type of antenna, and specifically expounds the latest research progress and design methods. As a collection, the series provides valuable resources to a wide audience in academia, the engineering research community, industry and anyone else who are looking to expand their knowledge of antenna methods. In addition, modern antenna series is also open. More antenna researchers are welcome to publish their new research results in this series.

More information about this series at http://www.springer.com/series/16786

Junping Geng · Chaofan Ren · Kun Wang ·
Han Zhou · Jing Zhang

Spoof Surface Plasmon Polaritons Antenna

Springer

Junping Geng
Department of Electronic Engineering
Shanghai Jiao Tong University
Shanghai, China

Kun Wang
Department of Electronic Engineering
Shanghai Jiao Tong University
Shanghai, China

Jing Zhang
Department of Electronic Engineering
Shanghai Jiao Tong University
Shanghai, China

Chaofan Ren
Department of Electronic Engineering
Shanghai Jiao Tong University
Shanghai, China

Han Zhou
Department of Electronic Engineering
Shanghai Jiao Tong University
Shanghai, China

Modern Antenna
ISBN 978-981-16-4720-8 ISBN 978-981-16-4721-5 (eBook)
https://doi.org/10.1007/978-981-16-4721-5

This Springer imprint is published by the registered company Springer Nature Singapore Pte Ltd.
The registered company address is: 152 Beach Road, #21-01/04 Gateway East, Singapore 189721, Singapore

Preface

With the continuous development of communication science and technology and the increasing demand of society, wireless communication technology is developed rapidly. As the bottleneck of wireless transmission system, antenna has also been payed much attention. Different antennas are used in different communication systems to satisfy different application environments and performance requirements.

Recently, spoof surface plasmon polaritons (SSPPs) have been proposed in microwave band in view of the surface plasmon polaritons (SPPs) in optics. Due to their small loss, deep subwavelength, tunable dispersion and conformality, SSPPs has unique advantages in miniaturization and high integration. SSPPs have been combined with other techniques to design a variety of excellent microwave RF devices, such as filters, transmission lines and power dividers. At the same time, the strong binding of SSPP structure to waves and the slow wave characteristics have also been applied to design the antenna. Antennas based on SSPPs have performed many advantages such as high gain, low profile, miniaturization and broadband. However, there are few books about SSPPs antenna principle and practical engineering applications, and there is no systematic theory and summary.

Based on the existing research of Prof. Junping Geng of Shanghai Jiao Tong University, this book analyzes the working mechanism of SSPPs and introduces the working principle and radiation characteristics of directional antenna, omnidirectional antenna, reconfigurable antenna and phase-mode antenna based on SSPPs structure.

The content of this book includes the following nine aspects:

Firstly, the research background and significance of SSPPs are expounded, and the current research status at home and abroad based on SSPPs is shown. The advantages of SSPPs are summarized.

Secondly, the basic principles of SSPPs structure such as dispersion relation, transmission mode and RCS characteristics are introduced.

Thirdly, compacted antenna with multi-bands are needed, especially in the cases such as mobile communications or radar systems. Consequently, the multipole antenna research based upon SSPPs is carried out.

Most of the antennas on the basis of SSPPs work in the even mode, and few studies report the odd mode. Based on the existing research, the SSPPs terminal antenna with asymmetric feeding is analyzed in Chap. 4 and a broadband and efficient terminal antenna is designed.

In applications such as routers, antennas are usually expected to have high gain, horizontal omnidirectional characteristics. Hence, in Chap. 5, three kinds of low-frequency omnidirectional SSPPs antennas are introduced.

Then, in Chap. 6, in order to solve the problem of polarization and non-circularity of antenna pattern, a novel rotating SSPPs structure (R-SSPPs) is proposed. By studying its basic electromagnetic characteristics and combining it with different feeding methods, an end-emitting circularly polarized antenna and a side-ray polarized omnidirectional antenna are designed.

Even though the single side arms SSPPs structure was mounted vertically on the ground to realize end-fire radiation in odd mode, its profile at lower frequency band is still relatively large for conformal design, especially to combine the SSPPs antenna into the shell to load bearing on high-speed vehicle. Therefore, in Chap. 7, a method to decrease the profile of the vertical polarized antenna with special-shaped SSPPs structure, such as T-shaped and m-shaped SSPPs structures, is introduced.

In Chap. 8, a large-angle pattern reconfigurable antenna based on SSPPs is designed and optimized. Through a reconfigurable circuit controlled by PIN diodes, this antenna realizes the switching between the odd mode and even mode of SSPPs, so as to achieve the purpose of switching the radiation beam from broadside lobe to end-fire lobe.

Finally, a novel phase-mode SSPPs antenna with dual port is introduced, which can work in odd mode, even mode and mixed modes with the characteristic of spatial lobe scanning simply by varying the phase difference between the two ports of one antenna element.

Shanghai, China Junping Geng

Acknowledgements The author would like to thank R. W. Ziolkowski, University of Technology Sydney, Global Big Data Technologies Centre, for several enlightening technical discussions. The author would also like to thank Prof. Ronghong Jin, University of Shanghai Jiao Tong Unversity, for multi-discussion to the contents. The author would also like to thank Dong Wang, Kaijie Zhuang, Chaofan Ren, Erwei Liu, Kun Wang, Han Zhou, Xuxu Cheng, Jiawei Han, Weinan Gao, Jing Zhang, Silei Yang, Jingzheng Lu, Da Su and Yangzhou Zhang for their support and assistance during the manuscript preparing.

Contents

List of Figures

List of Tables

Chapter 1
Introduction

Junping Geng

1.1 Background and Demand

Surface plasmon polaritons (SPPs) mainly exists in the frequency band above the far infrared. They are highly localized electron density waves propagating along the metal surface formed by the interaction of external electromagnetic field and free vibrating electrons on the metal surface. This kind of surface plasmon wave confines the electromagnetic energy in a small sub wavelength range and propagates forward, while the amplitude decreases exponentially in the direction perpendicular to the interface. Plasma has been applied in many fields, such as optoelectronics [1–5], nanomedicine [6–10], membrane technology [11–13], and SPPs have created a new research field of plasma, and quickly attracted the attention of many researchers. The plasma frequency of metal is in the infrared and optical frequency band, while in the microwave and millimeter wave frequency band, the skin depth of metal is very small, which shows as a perfect conductor rather than an electric plasma. Therefore, the research of SPPs in terahertz and microwave low frequency band has been in the blank.

In 2004, Pendry and his collaborators ever designed an artificial periodic dielectric hole array structure on the metal surface [14], which effectively confines electromagnetic waves on the surface, just like the SPPs wave in microwave band. It means they reduced the spoof plasma frequency on the artificial metal surface and opened the door to the research and application of spoof SPPs in the low frequency band. This structure with periodical "slot" and "hole" on the metal surface is called artificial surface plasmon polaritons (SSPPs) structure, or artificial SPPs waveguide (SSPPs waveguide), or spoof surface plasmon polaritons (SSPPs) structure. The surface electromagnetic wave supported by this periodical structure is called

J. Geng (✉)
Department of Electronic Engineering, Shanghai Jiao Tong University,
Shanghai 200240, China
e-mail: gengjunp@sjtu.edu.cn

J. Geng et al., *Spoof Surface Plasmon Polaritons Antenna*, Modern Antenna,
https://doi.org/10.1007/978-981-16-4721-5_1

artificial SPPs (SSPPs). SSPPs structure can flexibly adjust the dispersion characteristics of the artificial surface plasmons by controlling the geometric parameters of the metal surface, which is different from the natural surface plasmons. So, SSPPs structure can be used to design adjustable, reconfigurable and intelligent circuits, devices or antennas.

Strong binding, low loss and large wave number are the good characteristics of SSPPs, so that many conformal SSPPs structures are low bending low and easy processing. Therefore, the SSPPs antenna can be compacted with low prfofile and easy integration to work in broadband. In addition, due to the rapid attenuation of electromagnetic wave in the normal direction of the SSPPs structure, the antenna coupling can be reduced, which is suitable for the scene with narrow space. However, there are few researches on the antenna based on SSPPs at present, the form is relatively simple, and the performance is not diversified enough. To sum up, it is of great significance to explore various types, multi-functional and high-performance SSPPs antennas.

1.2 Research Development of the Surface Plasmon Polaritons and Spoof Surface Plasmon Polaritons

1.2.1 Research Progress on Surface Plasmons Polaritons

1.2.1.1 Surface Plasmons Polaritons

Assumed that the interface of metal/dielectric is on the *XOY* plane and *z*-axis is its normal direction. The metal surface is at z = 0. The semi infinite region of Z < 0 is filled by metal (assumes the permittivity of this region is ε_2), and upward half space of z > 0 is occupied by vacuum or dielectric material (the permittivity of this region is ε_1). The metal/dielectric material diagram is shown in Fig. 1.1. It shows that the metal/dielectric interface only supports TM mode, the wave number is $\boldsymbol{k}$, the angular frequency is ω. So, the electromagnetic field can be written as

$$\mathrm{E} = E_0 e^{-j(k_x x + k_z z - wt)} \tag{1.1}$$

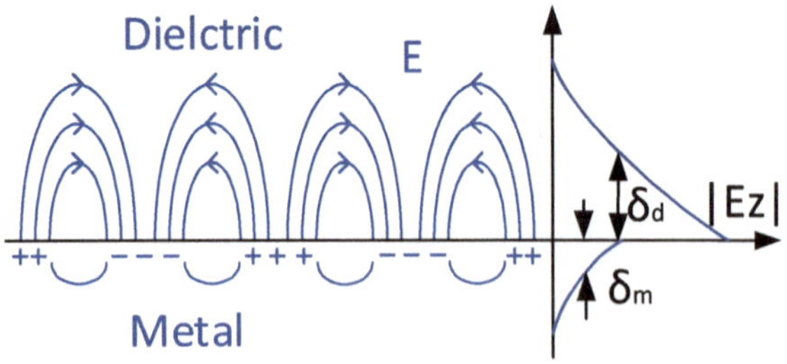

Fig. 1.1 Metal/dielectric material diagram

Based on the Ampere's law, $\nabla \times \vec{H} = ik_0\varepsilon\vec{E}$, and the boundary conditions of the metal/dielectric interface, we can get the equations at Z = 0:

$$\frac{k_{z_1}}{\varepsilon_1} + \frac{k_{z_2}}{\varepsilon_2} = 0 \tag{1.2}$$

$$k_x^2 + k_{zi}^2 =_i \left(\frac{\omega}{c}\right)^2 \tag{1.3}$$

In the visible band, the permittivity of the metal is not infinite, but dispersive, which is closely bound up with the wavelength of the incident light wave. In another side, the loss of the metal in optical band is serous and can't be ignored, so that the permittivity is set as a complex number $\varepsilon_1 = \varepsilon_r' + j\varepsilon_r''$, and the corresponding wave number of SPPs is set as complex wave number too, assuming $k_x = k' + ik''$,

$$k' = \frac{w}{c}\left(\frac{\varepsilon_r'\varepsilon_2}{\varepsilon_r' + \varepsilon_2}\right)^{1/2} \tag{1.4}$$

$$k'' = \frac{w}{c}\left(\frac{\varepsilon_r'\varepsilon_2}{\varepsilon_r' + \varepsilon_2}\right)^{3/2} \frac{\varepsilon_r''}{2(\varepsilon')^2} \tag{1.5}$$

$$k_{SPPs} = k_0\sqrt{\frac{\varepsilon_2\varepsilon_1}{\varepsilon_2 + \varepsilon_1}} \tag{1.6}$$

Thees formulas are just the dispersive functions of SPPs, and Fig. 1.2 showes the dispersive curve under the light line.

1.2.1.2 The Development of SPPs

In the optical band, the metal layer has a big impact on the surface plasmon propagation on the metal surface. With the new nanotechnology and

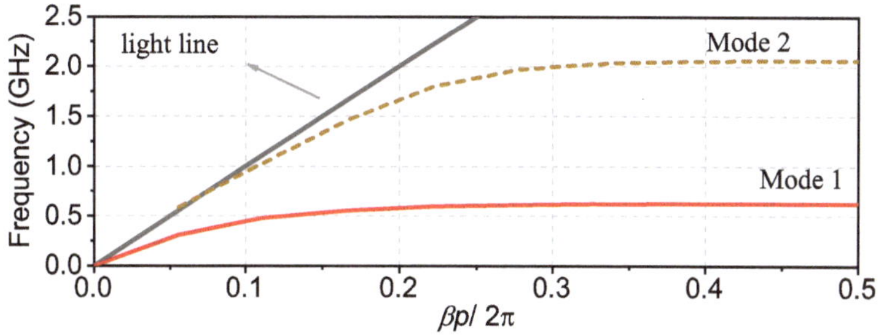

Fig. 1.2 The dispersive curve of SPPs

micromachining technology developing quickly, the new theory and technologies based on SPPs have been the hotspot since 1990s, and absorbed many reseachers. In [15], Lezec et al. prresented the metal film with a series of circular structures around the holes. It shows that transmission enhancement is not only realized, but also the transmission beam is obtained with the emission angle less than 3°, and the beam width is approximately equal to the wavelength. This constraint can be used to light emission of laser diode and semiconductor laser to realize higher quality optical devices.

SPPs can also be applied to nano lithography. By using silver as the lens material, Fang et al. obtained an image of sub diffraction limit, which is 1/6 of the incident wavelength in [16]. It is noted that the resolution of SPPs is raised from 320 to 89 nm, and it is about 4 times higher than that of ordinary etching technology.

Surface plasmon polaritons can also be used to enhance surface Raman scattering. Baumberg et al. realized the enhancement of Raman scattering by surface plasmon polaritons by constructing nanostructures on the metal surface [17]. The nanostructures on the metal surface solved the strict requirements of SERS on the angle, making it more convenient to excite SERS. SERS technology has been gradually applied in the fields of surface science and electrochemistry. It has been widely used in single molecule detection, chemical industry, biomedicine, nano materials and sensors. Hope et al. Used SERS to analyze the interaction between copper substrate and butyl acetyl carbonyl thiourea, and proposed that butyl acetyl carbonyl thiourea can be introduced into Cytec industry for flotation separation of copper sulfide ore [18]. Viets et al. used hydrofluoric acid to etch the optical fiber, and made a prototype optical fiber sensor based on sets [19].

In recent years, the application of surface plasmon polaritons in solar cells has been absorbed more and more attention. In the research of solar cells, we usually make a thick enough active layer to absorb sunlight and form photogenerated carriers to collect current. In 2007, Green et al. realized the light absorption efficiency of silicon-based solar cells with different structures by using silver nanoparticles [20]. They proposed that the improvement of solar absorption efficiency by surface plasmon is not limited to a certain wavelength, but covers the whole solar spectrum.

The most commonly used surface plasmon polaritons sensor structures are prism structure, grating structure and waveguide structure [21].

To sum up, the SPPs has been widely used in many fields due to its sub wavelength constraint, localization enhancement and dispersion characteristics. It is not limited in the fields mentioned above. Many current research hotspots, such as invisibility cloak and metamaterials, are closely related to surface plasmon in principle, although the surface plasmon polaritons is equidistant at present, the progeny has been well developed. Only because there are many aspects in the theoretical mechanism that are worth exploring, and with the continuous expansion of the application field, we believe that the surface plasmon polaritons has a very broad research and development prospect, whether in the depth of theory or in the expansion of practical application.

1.2.1.3 SPPs Device

In 2005, Berini et al. proposed the transmission of surface plasmon polaritons wave by embedding a metal film with a preferential width in the medium, which can achieve 13.6 mm transmission with the incident wavelength 1.55 μm, but its transverse beam width constraint is several microns [22]. In order to achieve subwavelength constraint of transverse beam width, the width of metal film can be narrowed. Krenn et al. used 800 nm wavelength light to excite SPP, and realized several micron transmission distance on 200 nm wide and 50 nm thick metal nanowires [23]. The reason is that the resistance of nanowires is relatively large, and the thermal loss limits the propagation distance of SPP. In order to reduce the ohmic loss, the structure of metal nanoparticles can be used to excite SPP to achieve longer propagation distance.

In order to realize the subwavelength constraint of the light field in the transverse and longitudinal directions, Bozhevolnyi et al. proposed the V-groove surface plasmon waveguide [24]. They fabricated the V-groove waveguide by using the concentrated ion beam etching method, and coupled the light into the waveguide by using the tapered fiber alignment method to realize long-distance transmission.

The first is passive devices such as coupler and beam splitter based on surface plasmon polaritons waveguide. Bozhevolnyi et al. Proposed the V-shaped waveguide structure and applied it to the design of photonic devices, and proposed the Y-beam splitter, interferometer, ring resonator and wavelength division multiplexer based on the V-shaped groove structure [25, 26]. The insertion loss of ring resonator based on surface plasmon is less than 3 dB, and its area is only 200 μm^2, which is equivalent to $100\lambda^2$. Yin et al. Proposed SPP coupler based on nanoelectricity [27].

Compared with waveguides and passive devices, the design of active devices based on surface plasmon polaritons is more responsible and plays the most important role in photonic integration. Krasavin et al. proposed a variety of surface plasmon polaritons based active structures [28]. They use grating or prism structure to excite SPPs, and use the temperature sensitive or photosensitive characteristics of polymer film to realize switching operation. Bozhevolnyi et al. proposed a Mach–Zehnder modulator based on SPPs by connecting the metal strip in the loaded SPPs waveguide with the external electrode and changing the refractive index of the dielectric layer through the thermo optic effect after loading the signal [29].

In addition to optical switches and modulators, SPPs is also gradually introduced into the research of lasers and photodetectors. Semiconductor lasers based on InGaN face the problem of luminous efficiency. This is because most of the power is lost to the air, and the enhancement characteristics of SPPs is used. The addition of SPPs will greatly improve the spontaneous emission efficiency. After the introduction of SPPs by Okamoto et al., the luminescence efficiency of InGaN/GaN quantum well at 440 nm is enhanced by 32 times [30]. In addition, if a metal nano aperture is added to the top of the GaAs vertical cavity surface emitting laser, the luminous efficiency can be enhanced by 1.8 times. The signal voltage can be doubled [31].

Berini et al. proposed a schottky structure surface plasmon photodetector [32]. By crossing a layer of metal film on the silicon substrate, the transmission of SPP is realized. At the same time, the electrode voltage of the Schottky diode is formed, and the power of the incident light can be obtained. The minimum loss of SPP is 1.204 db/μm.

1.2.2 Spoof Surface Plasmon Polaritons

For a long time, the research of surface plasmon polaritons is mainly limited in the optical band or higher frequency range. This is because in the optical frequency or higher frequency range, the metal presents a negative dielectric constant, while in the microwave frequency range, the real part and imaginary part of the dielectric constant of the metal are very large, and the value k_{SPPs} is close to the wave vector in the medium, resulting in a very small skin depth in this frequency band, indicating that there is little field penetrating into the conductor.

Although metals in terahertz or microwave band can not support highly constrained surface plasmons in natural form, ideal good conductors with specific surface structure can support surface electromagnetic waves similar to surface plasmons.

In 2004, Pendry and his collaborators in of Imperial College of Technology designed an artificial periodic dielectric hole array structure on the metal surface [33]. It is theoretically proved that when electromagnetic wave propagates along this structure, it will be bound to the surface, forming surface waves similar to SPPs waves, thus opening the door of SPPs research in the microwave field.

The artificial surface structure proposed by Pendry et al. has good characteristics, but it adopts a three-dimensional structure with large size. In order to meet the needs of miniaturization and low profile, scholars began to study the artificial surface plasmon polaritons of planar guided wave structure.

In 2013, Cui Tiejun proposed an artificial surface plasmon polaritons structure conformal to flexible materials [34], as shown in Fig. 1.2. The folded metal strip is printed on a flexible ultra-thin dielectric substrate with a width of sub wavelength and a thickness of a few hundredths of the working wavelength. The electric field amplitude of this structure decays exponentially in two dimensions perpendicular to the transmission direction of electromagnetic surface wave, realizing subwavelength bound transmission [35]. Professor Cui Tiejun and others have realized the conformal of the artificial surface plasmon, which enables the electromagnetic surface wave to transmit efficiently along any curved surface.

Based on the research of transmission line, filter, power divider and other microwave devices [36–38], researchers began to apply SSPPs structure to antenna design, and achieved certain results. The main types of SSPPs antenna will be introduced below.

1.2.3 Antenna Fed by Spoof Surface Plasmon Polaritons Waveguide

Amin Kianinejad et al. proposed using SSPPs to feed the dielectric resonant antenna in 2016, and verified it through experiments [39]. The SSPPs transmission line has no ground, which reduces the influence of the floor on the dielectric resonance antenna. Through a bent SSPPs structure, the antenna can realize the excitation of the horizontal polarization mode of the dielectric resonator.

Guan et al. used SSPPs waveguide structure to excite a circularly polarized array in 2017 [40]. The antenna array consists of a series of circular patches, and the feeding structure of SSPPs is located under the dielectric layer. The simulation and measurement results also show that the scanning range of the array can be effectively improved by using the slow wave characteristics of SSPPs feed structure.

1.2.4 Leaky Wave Antennas Based on Spoof Surface Plasmon Polaritons

Due to the strong binding effect of SSPPs structure on electromagnetic wave, it is widely used in transmission line design. Some people introduce periodic modulation unit [41] or coupling unit [42] to transform the bound electromagnetic field in SSPPs transmission line into radiated electromagnetic wave in free space.

The periodically modulated SSPPs frequency scanning antenna is proposed in Ref. [41]. The antenna can make the electromagnetic wave in SSPPs waveguide radiate into free space by periodically inserting rhombic modulation elements. The beam direction can be adjusted by adjusting the layout interval.

The near-field electric field distribution of the antenna is obvious that the electromagnetic wave is continuously radiated into the free space in the process of SSPPs waveguide transmission.

A frequency scanning antenna based on SSPPs is implemented in Ref. [41] by modulation method. On this basis, some scholars proposed an improved antenna form to increase the frequency scanning angle of SSPPs antenna [43]. The antenna adopts structure VI to realize the standing wave distribution on the transmission line of SSPPs, and by periodically placing circular patches on both sides of the transmission line of structure III, the electromagnetic wave of SSPPs mode is radiated into the free space through coupling. However, this kind of antenna is based on a dual port transmission line, and the beam in the end fire direction cannot be realized.

1.2.5 Endfire Antenna Based on Spoof Surface Plasmon Polaritons

A high efficiency traveling wave antenna based on SSPPs is proposed in Ref. [44]. Due to the unilateral SSPPs structure, the antenna profile is low. The antenna is fed by coplanar waveguide, and SSPPs waveguide is asymmetric structure with slots only on one side. The antenna has good end fire effect in 7.5–8.5 GHz. The antenna size is 0.32×0.1, the maximum efficiency is 96%, and the end fire bandwidth is 12.5%.

1.2.6 Spoof Surface Plasmon Polaritons Antenna with Parasitic Structure

In Ref. [45], the SSPPs structure is placed near the traditional monopole for coupling feeding, and the omni-directional electromagnetic wave is converted into directional radiation by using the characteristics of high binding and efficient transmission of traveling wave of SSPPs. The end fire of this kind of antenna is different from that of coplanar waveguide fed end fire antenna described in Sect. 1.2.3.

A monopole feeds the SSPPs structure, and a double-layer dielectric plate clamps the SSPPs structure to improve the binding capacity of SSPPs. The gradual structure at the feed end helps to couple the electromagnetic energy into the SSPPs structure, and the gradual structure at the end helps to radiate the electromagnetic wave and realize the end shooting. The simulation results show that the antenna has a 9.7 dbi end transmit gain at 10 GHz. The impedance bandwidth of the feed monopole is limited, so the impedance bandwidth of the whole antenna is only 15%.

1.3 Development Trend and Problems

In recent years, artificial surface plasmon transmission lines with sub wavelength field localization, low loss, conformal transmission and tunable dispersion characteristics have also attracted extensive attention. They are considered as the most effective new transmission lines to solve the bottleneck of microwave and millimeter wave integrated circuit technology.

SSPP antenna provides a new idea for antenna research, especially in the aspects of multipole radiation, miniaturization, multi-dimensional beam control and so on, which has great potential. At the same time, it is also convenient to realize filter antenna.

1.4 Book Arrangement

Focusing on the principle, technology and application of SSPPs antenna, this book has nine chapters. The specific contents are as follows:

In this chapter, the application background and significance of the antenna are described, the research status of artificial surface plasmon at home and abroad is introduced, and the main work of this paper is briefly introduced.

In Chap. 2, the basic principle and simulation method are introduced. Firstly, the basic principle of artificial surface plasmon is described, including dispersion relation and transmission mode of artificial surface plasmon. Then the antenna based on artificial surface plasmon is introduced. Finally, the simulation methods used in the antenna design are briefly introduced, including dispersion curve simulation and RCS simulation.

In Chap. 3, a multi band antenna based on SSPP theory is introduced. By discussing the performance changes brought by applying SSPP theory to fish antenna, the specific parameters affecting the antenna are found according to the change rules, and finally the multi band antenna based on SSPP is realized. Finally, through the processing and testing of the antenna to explore the actual performance of the antenna, the simulation and measurement results are basically consistent, and the designed antenna has good performance.

In Chap. 4, two end fire antennas based on artificial surface plasmons are introduced. Firstly, a coplanar waveguide fed end fire antenna based on artificial surface plasmon structure is introduced, and the transmission mode and radiation mode of artificial surface plasmon are analyzed. Then, a wideband end fire antenna based on artificial surface plasmon is introduced, the influence of parameters is analyzed, and the real object is processed and tested.

In Chap. 5, a low profile antenna based on artificial surface plasmon is proposed. Firstly, the transmission mode and pattern of the traditional artificial surface plasmon end fire and side fire antennas are introduced. Then, a reconfigurable feed network is designed, based on which the antenna model is designed. The influence of several important parameters on the antenna performance is analyzed. Finally, the processing and testing of the antenna are completed.

In Chap. 6, combined with the specific application scenarios, a low-frequency antenna based on SSPPs is proposed by using the new SSPP theory. By discussing the application of SSPP theory and the performance changes brought by the antenna, the specific parameters affecting the antenna are found according to the changing rules, and finally the miniaturized low profile broadband high efficiency antenna based on SSPP theory is realized. Finally, through the processing and testing of the antenna to explore the actual performance of the antenna, the simulation and measurement results are basically consistent, and the designed antenna has good performance.

In Chap. 7, a novel rotating artificial surface plasmon structure is introduced, and compared with the traditional SSPPs structure, the dispersion relation and odd and even mode characteristics are studied. On this basis, a circularly polarized end fire antenna and a linearly polarized side fire omni-directional antenna are designed. The influence of several key parameters of the results on the antenna performance is analyzed, and the parameter optimization is completed. Finally, the physical processing and testing are carried out. The final experimental results are basically consistent with the simulation results, and the two antennas have good radiation performance.

Chapter 8 mainly introduces the SSPPs mode reconstruction technology based on PIN diode. The switch circuit is constructed by pin diode to realize the odd mode and even mode switching, and the wide beam side shot and end shot switching is realized correspondingly.

In Chap. 9, the SSPPs phase mode antenna excited by two ports is introduced. The excitation principle and radiation variation law of odd mode, even mode and mixed mode of SSPPs antenna are introduced.

References

1. J.D. Kraus, R.J. Marhefka et al., *Antenna*, 2nd edn. (Tata McGraw-Hill, 2001)
2. D. Yaolin, *Design and Simulation of L-Band High Power Radial Line Array Antenna* (Southwestern Jiao Tong University, 2009)
3. M.B. Diaz, I. Campillo, J.S. Dolado et al., Dual-band low-profile corrugated feeder antenna. IEEE Trans. Antennas Propag. **54**(2), 340–350 (2006)
4. W. Aiting, G. Boran, Broadband microstrip patch antenna using a superstrate layer. J. Hangzhou Instit. Electron. Technol. **24**(6), 4–7 (2004)
5. W. Yadong, *The Research of Novel High Gain antenna Based on Transmission Enhancement* (Anhui University, 2017)
6. A.D. Yaghjian, in *Increasing the Supergain of Electrically Small Antennas Using Metamaterials*. European Conference on Antennas and Propagation, IEEE, 2009, pp. 858–860
7. S.R. Best, in *Progress in the Design and Realization of an Electrically Small Huygens Source*. International Workshop on Antenna Technology, IEEE, 2010, pp. 1–4
8. J. Peng, R.W. Ziolkowski, Metamaterial-inspired, electrically small huygens sources. IEEE Antennas Wirel. Propag. Lett. **9**(1), 501–505 (2015)
9. R.W. Ziolkowski, Low profile, broadside radiating, electrically small huygens source antennas. IEEE Access **3**, 2644–2651 (2017)
10. T. Mingchun, W. Hao, S. Ting et al., *Electrically Small Plane Huygens Source Antenna*, CN 106252861 A, 2016
11. R.W. Ziolkowski, Using huygens multipole arrays to realize unidirectional needle-like radiation. Phys. Rev. **X 7**, 031017 (2017)
12. T. Wenxuan et al., Artificial surface plasmon and its application in microwave band. Acta Electron. Sin. **39**(1), 231–239 (2017)
13. K. Gusheng, *Radiation Regulation of Artificial Surface Plasmons* (Southeast University, 2016)
14. L. Bo, *Research on Surface Plasmon and Surface Wave Plasma of Microwave Plasma* (Fudan University, 2006)
15. H.J. Lezec, A. Degiron, E. Devaux, et al. Beaming light from a subwavelength aperture. Science 297 (5582), 820 (2002)
16. N. Fang, H. Lee, C. Sun et al., Sub–diffraction-limited optical imaging with a silver superlens. Science **308**(5721), 534–537 (2005)

17. J.J. Baumberg, T.A. Kelf, Y. Sugawara et al., Angle-resolved surface-enhanced Raman scattering on metallic nanostructured plasmonic crystals. Nano Lett. **5**(11), 2262 (2005)
18. G.A. Hope, R. Woods, S.E. Boyd, et al. A SERS spectroelectrochemical investigation of the interaction of butylethoxycarbonylthiourea with copper surfaces. Colloids Surf., A **232**(2), 129–137 (2004)
19. C. Viets, W. Hill. Fibre-optic SERS sensors with angled tips. J Mol Struct. **565**(2), 515–518 (2001)
20. S. Pillai, K.R. Catchpole, T. Trupke, et al., Surface plasmon enhanced silicon solar cells. J. Appl. Phys. **101**(9), 093105–093105–8 (2007)
21. W. Jingquan, L. Huimin, S. Sha et al., Theoretical analysis of interference nanolithography of surface plasmon polaritons without a match layer. China Physics Express English Edition **26** (8), 155–157 (2009)
22. P. Berini, R. Charbonneau, N. Lahoud et al., Characterization of long-range surface-plasmon-polariton waveguides. J Appl Phys. **98**(4), 10484 (2005)
23. J.R. Krenn, J.C. Weeber, Surface plasmon polaritons in metal stripes and wires. Philos Trans Math Phys Eng Sci. **362**(1817), 739–756 (2004)
24. S.I. Bozhevolnyi, V.S. Volkov, E. Devaux et al., Channel plasmon-polariton guiding by subwavelength metal grooves. Phys Rev Lett. **95**(4), 046802 (2005)
25. V.S. Volkov, S.I. Bozhevolnyi, E. Devaux et al., Wavelength selective nanophotonic components utilizing channel plasmon polaritons. Nano Lett. **7**(4), 880 (2007)
26. S.I. Bozhevolnyi, V.S. Volkov, E. Devaux et al., Channel plasmon subwavelength waveguide components including interferometers and ring resonators. Nature. **440**(7083), 508 (2006)
27. L. Yin, V.K. Vlasko-Vlasov, J. Pearson et al. Subwavelength focusing and guiding of surface plasmons. Nano Lett. **5**(7), 1399 (2005)
28. A.V. Krasavin, A.V. Zayats, N.I. Zheludev, Active control of surface plasmon-polariton waves. J Opt A: Pure Appl Opt. **7**(2), S85 (2005)
29. J. Gosciniak, S.I. Bozhevolnyi, Performance of thermo-optic components based on dielectric-loaded surface plasmon polariton waveguides. Sci Rep. **3**(3), 1803 (2013)
30. K. Okamoto, I. Niki, A. Scherer et al., Surface plasmon enhanced spontaneous emission rate of Ingan/Gan quantum wells probed by time-resolved photoluminescence spectroscopy. Appl Phys Lett. **87**(7), 1687 (2005)
31. J. Hashizume, F. Koyama, Plasmon enhanced optical near-field probing of metal nanoaperture surface emitting laser. Opt Express. **12**(25), 6391–6 (2004)
32. A. Akbari, R.N. Tait, P. Berini, Surface plasmon waveguide Schottky detector. Opt Express. **18**(8), 8505 (2010)
33. J.J. Xu, J.Y. Yin, H.C. Zhang et al., Compact feeding network for array radiations of spoof surface plasmon polaritons. Sci. Rep. **6**, 22692 (2016)
34. S. Shiwei, *FDTD simulation of terahertz wave propagation in subwavelength metal structures* (Shanghai University, 2010)
35. J.B. Pendry, L. Martín-Moreno, F.J. Garcia-Vidal, Mimicking surface plasmons with structured surfaces. Science **305**(5685), 847 (2004)
36. Z. Wenjuan, *Theory and Application of Microstrip Artificial Surface Plasmon* (University of Science and Technology of China, 2016)
37. A.P. Hibbins, B.R. Evans, J.R. Sambles, Experimental verification of designer surface plasmons. Science **308**(5722), 670 (2005)
38. N. Pingping, *Field Enhancement and Crosstalk Suppression Based on Artificial Surface Plasmons*. Nanjing University of Aeronautics and Astronautics, 2016
39. K. Wang, D.M. Mittleman, Metal wires for terahertz wave guiding. Nature **432**(7015), 376 (2004)
40. A.I. Fernándezdomínguez, E. Moreno, L. Martínmoreno et al., Guiding terahertz waves along subwavelength channels. Phys. Rev. B Condensed Matter **79**(23), 233104 (2009)

41. S. K. Gu, H. F. Ma, B. G. Cai, T. J. Cui, Continuous leaky-wave scanning using periodically modulated spoof plasmonic waveguide. Sci Rep. **6**, 29600, 2016
42. Q. Zhang, Q. Zhang, Y. Chen, Spoof surface plasmon polariton leaky-wave antennas using periodically loaded patches above PEC and AMC ground planes. IEEE Antennas &Wireless Propagation Letters, vol. 16, pp. 3014–3017, 2017
43. D. Liao, Y. Zhang, H. G. Wang. Wide-angle frequency-controlled beam scanning antenna fed by standing wave based on the cut-off characteristics of spoof surface plasmon polaritons. IEEE Antennas & Wireless Propagation Letters
44. A. Kandwal, Q. Zhang, X. Tang, L. W. Liu, G. Zhang, Low-profile spoof surface plasmon polaritons traveling-wave antenna for endfire radiation. IEEE Antennas & Wireless Propagation Letters, vol. 17, no. 2, pp. 184–187, 2017
45. Y. J. Han, J. Zhang, Y. Li, J. Wang, S. Qu, A. Zhang, 360° scanning multi-beam antenna based on spoof surface plasmon polaritons. Acta Physica Sinica, vol. 65, no.14, 2016

Chapter 2
Basic Principles of Spoof Surface Plasmon Polaritons

Junping Geng

Surface plasmon polaritons (SPPs) is a kind of surface electromagnetic wave mode formed by the interaction of metal surface charge and electromagnetic wave in the medium. This effect is mainly in the optical frequency band. Later, through the equivalent effect of this rough surface structure, it was introduced into millimeter wave, even microwave band, and used in the design of devices and antennas, which is called spoof surface plasmon polaritons (SSPPs).

2.1 Basic Principle of Surface Plasmon Polaritons

To SPPs wave, the interface between the metal and the dielectric, the real part of the dielectric constant of the metal is negative in the frequency range below the plasma frequency; on the other side, the dielectric constant of the dielectric material is positive. According to the boundary conditions of the electromagnetic field, in order to ensure the continuity of the normal component of the electric displacement vector, the surface polarized charge will appear on the interface. The surface charge oscillates collectively under the action of the electromagnetic field with the same resonance frequency, showing unique electromagnetic characteristics: it propagates along the interface between conductor and medium, and decays rapidly along the vertical direction of the interface [1]. Therefore, the surface plasmon is confined to the metal surface and enhances the electric field on the metal surface.

J. Geng (✉)
Department of Electronic Engineering, Shanghai Jiao Tong University,
Shanghai 200240, China
e-mail: gengjunp@sjtu.edu.cn

J. Geng et al., *Spoof Surface Plasmon Polaritons Antenna*, Modern Antenna,
https://doi.org/10.1007/978-981-16-4721-5_2

2.1.1 Dispersive Material Models

With any lossy real or artificial material, dispersive effects must be taken into account. This is especially true of all epsilon negative (ENG), mu-negative (MNG), and double negative (DNG) and double positive (DPS) metamaterials. The lack of general purpose dispersion models was a drawback of former versions of CST. However, in recent versions, CST has added some functionality that allows one to model frequency dependent, passive negative permittivity and permeability materials. Unfortunately, there are some difficulties yet in the time domain solver for negative conductivity or for obtaining the desired sign of the imaginary part of the refractive index to model active media. Nonetheless, we have found [2] that a user can introduce gain in the frequency domain solver by manipulating the material parameters in the "history list" manually. It is for this reason that this manual process is introduced below to analyze the active CNP problem with the CST software tools. We have confirmed that the CST results obtained in this manner reproduce the analytically obtained ones for the canonical passive and active spherical CNPs.

2.1.1.1 Optical Material Properties

In [3], an improved model of the metal plasmonic shells at optical wavelengths between 200 and 1800 nm was presented, and the empirical values given in [4, 5] were reproduced by the improved model. The permittivity can be written as two parts, a size dependent Drude response and an interband transition response.

$$\varepsilon(\omega, R) = \varepsilon_{Drude}(\omega, R) + \chi_{Interband}(\omega) \tag{2.1}$$

Here, R is the thickness of the metal shell and the Drude permittivity is can be described as

$$\varepsilon_{Drude}(\omega, R) = 1 - \frac{\omega_p^2}{\Gamma(R)^2 + \omega^2} + j\frac{\Gamma(R)^2\omega_p^2}{\omega\left(\Gamma(R)^2 + \omega^2\right)} \tag{2.2}$$

where ω_p is the plasma frequency and Γ is the collision frequency. When the size of the material approaches and becomes smaller than the bulk mean free path length of the conduction electrons in it, i.e., when it is thinner than 100 nm, the size influence must be considered. It is treated as an alteration in the mean free path which is then incorporated into the Drude model through the size dependent damping frequency. The damping frequency is assumed to be:

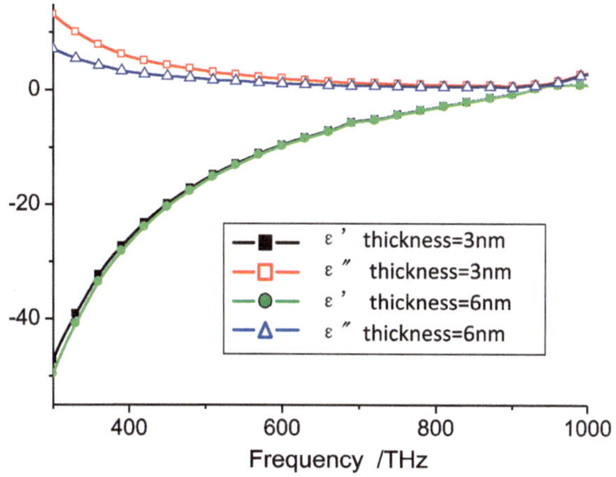

Fig. 2.1 Permittivity of silver films whose thicknesses are 3 and 6 nm. Figure reproduced with permission from Ref. [6], © 2011 IEEE

$$\Gamma(R) = \Gamma_{\infty} + \frac{AV_F}{R} \tag{2.3}$$

Here A is a constant, and is assumed to be approximately unity, i.e., A ~ 1. V_F is the Fermi velocity. The Drude parameters and the Fermi velocity values for silver in our simulations were refered from [3]: m*/m = 0.96, N = 5.85 × 10^{28} m^{-3}, ω_p = 1.39269 × 10^{13} s^{-1}, V_F = 1.39 × 10^6 m/s, and A = 1.0. For example, when the silver films thickness is 3 and 6 nm, Using the above process, the permittivities of silver films were calculated, which are shown in Fig. 2.1.

2.1.1.2 Active Media in Nano-Shells

To further study the SPP characteristics, the silver film coated nano particles (CNPs) model were constructed. The core of the passive CNPs models has been filled with silicon dioxide, SiO_2. Because SiO_2 has very low loss characteristics in microwave band, and we set the relative permittivity to be a constant $\varepsilon_r = 2.05$, yielding the refractive index $n = \sqrt{\varepsilon_r} = 1.432$. However, if the filling media is a higher-loss dielectric, the dielectric constant will then include an imaginary part, like Eq. (2.4)

$$\varepsilon(\omega) = \varepsilon'(\omega) - j\varepsilon''(\omega) \tag{2.4}$$

then I the refractive index expression contains an imaginary term, κ for the loss material too. The relation between the permittivity and the real and imaginary parts of the refractive index can be expressed by the expression:

$$\varepsilon_r = (n - j\kappa)^2 = n^2 - \kappa^2 - j2n\kappa \tag{2.5}$$

where, κ is varied over a range of values that mean the expected loss or gain in the core of the nano-shell, where κ > 0, = 0, and < 0 are just for a lossy, lossless, and gain medium, respectively.

2.1.1.3 Lorentz Model

The resonance behavior of a material can be described by the *Lorentz model* [7], which describes the permittivity as

$$\varepsilon_r = (\omega) = \varepsilon_\infty + \frac{(\varepsilon_S - \varepsilon_\infty)\omega_0^2}{\omega_0^2 + j\omega\Gamma - \omega^2} \tag{2.6}$$

For the Er^{3+} doped SiO_2, we can set $n = \sqrt{\varepsilon_r} = 1.432$, $\kappa = -0.25$ [8, 9] at the resonance frequency of the structure, for example at $f_0 = 600$ THz. Then at $f = f_0$, $\omega = \omega_0 = 2\pi f_0$, Eq. (2.6) becomes

$$\varepsilon_r(\omega_0 = 2\pi f_0) = \varepsilon_\infty + \frac{(\varepsilon_s - \varepsilon_\infty)\omega_0}{j\Gamma} \tag{2.7}$$

Combining Eqs. (2.3), (2.8) and (2.10), we obtain $\varepsilon_\infty = 1.9875$ and $\frac{(\varepsilon_s - \varepsilon_\infty)\omega_0}{\Gamma} = 2n\kappa$, so that

$$\varepsilon_s = \frac{2n\kappa\Gamma}{\omega_0} + \varepsilon_\infty = \frac{2n\kappa\Gamma}{\omega_0} + 1.9875 \tag{2.8}$$

As is well known, Γ is the collision frequency (or damping factor). It is frequently related to the resonant frequency for simplicity of the model. It is set in the range $(10^{-3}–10^{-1})\ \omega_0$, to represent low to high loss materials. Some values of Γ and ε_s are listed in Table 2.1 for reference.

Based on the above derivation and values of its constants, the real and imaginary parts of the Lorentz model of the permittivity of the doped SiO_2 are given in Fig. 2.2. The curves exhibit typical behavior when gain is present in a Lorentz medium. In the following cases, the lower loss value $\Gamma = 10^{-3}\ \omega_0$ is selected.

Table 2.1 Relationship between the static permittivity and the collision frequency

Γ	$\varepsilon_S = 2n\kappa\Gamma/\omega_0 + \varepsilon_\infty$	
$10^{-3}\ \omega_0$	1.98678	$n = 1.432$
$10^{-2}\ \omega_0$	1.98034	$\kappa = -0.25$
$10^{-1.3}\ \omega_0$	1.95162	$f_0 = 600$ THz, $\omega_0 = 2\pi f_0$
$10^{-1}\ \omega_0$	1.91591	

Table reproduced with permission from Ref. [6], © 2011 IEEE

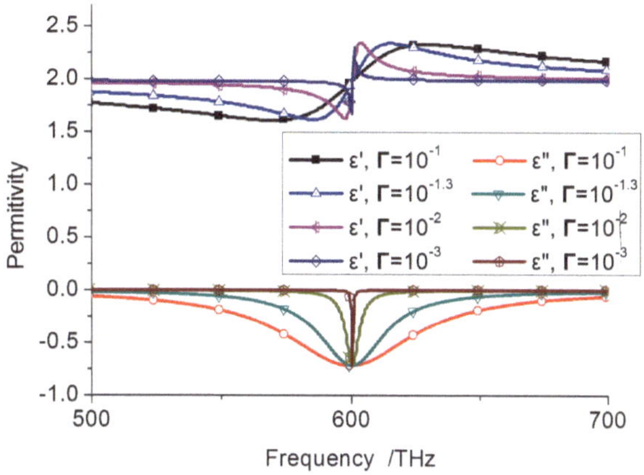

Fig. 2.2 Permittivity of doped SiO_2 with different Γ at $\omega_0 = 120\pi \times 1012$. Figure reproduced with permission from Ref. [6], © 2011 IEEE

2.1.2 *Plane Wave Excitation Performance Terms*

Scattering cross section (SCS) and the absorption cross section (ACS) are often applied to characterize the behavior of a structure being excited by an electromagnetic plane wave. The SCS is defined as the total integrated scattered field power normalized by the irradiance of the incident field. The ACS is defined by the net flux through a surface surrounding the structure normalized by the incident field irradiance. Both the ACS and SCS can be calculated by Poynting's theorem through the power scattered and absorbed by the scatter. They are given by the following expressions:

$$P_{scat} = \mathrm{Re}\left\{\frac{1}{2}\iint_S [\vec{E}_S \times \vec{H}_S^*] \cdot \hat{n} dS\right\} \tag{2.9}$$

$$P_{ads} = \mathrm{Re}\left\{\frac{1}{2}\iint_S [\vec{E}_{tot} \times \vec{H}_{tot}^*] \cdot \hat{n} dS\right\} \tag{2.10}$$

where S is a integration surface that completely surrounds the particle and $\hat{n}$ is the unit vector outward pointing normal to it. Labeling the incident, scattered, and total fields by the subscripts *inc*, *scat*, and *tot*, respectively. The total SCS, ACS and extinction cross-section (ECS) are thus defined from the ratio of the scattered or absorbed power to the incident irradiance I_{inc} as following:

$$\sigma_{scat} = \frac{P_{scat}}{I_{inc}} \tag{2.11}$$

$$\sigma_{abs} = \frac{P_{abs}}{I_{inc}} \tag{2.12}$$

$$\sigma_{ext} = \sigma_{scat} + \sigma_{abs} \tag{2.13}$$

These quantities can be calculated by the CST post-processing tools.

2.1.3 Radiated Power by Small Current

Similarly, to characterize the behavior of a structure being excited by an EHD, the total radiated poweris introduced by the Poynting's vector-based expression:

$$P_{tot} = \lim_{r\to\infty} \frac{1}{2} \int_{\theta=0}^{\pi} \int_{\varphi=0}^{2\pi} \mathrm{Re}\{\vec{E}_{tot}(r,\theta,\varphi) \times \vec{H}^{*}_{tot}(r,\theta,\varphi)\}\hat{n}dS \tag{2.14}$$

where the intergration surface S is sphere enclosing the entire nano-antenna system. For an EHD with a dipole current moment: $p_s = I_0 d$, radiating in free space, the radiated power can be expressed as [7, p. 155]

$$P_i = \frac{\eta_0 \pi}{3} \left| \frac{p_s k_0}{2\pi} \right|^2 \tag{2.15}$$

Here, $\eta_0 = 376.7\ \Omega$ is the characteristic impedance of free space. The radiated power ratio (PR) by the scatter excited by EHD is then defined simply as:

$$PR = \frac{P_{tot}}{P_i} \tag{2.16}$$

2.1.4 Passive CNP Model

As in [9], the CNP is filled with SiO_2 with a permittivity equal to 2.05. The radius of the core is taken to be 24 nm. The coated film is silver and has a thickness 6 nm. The source is the plane wave: $\vec{E} = \hat{z}E_0 \cos(\omega t - kx)$. Both the time domain and frequency domain solvers are selected to analyze the passive CNP. The CST geometry is shown in Fig. 2.3 and the predicted ACS and SCS values are given in Figs. 2.4 and 2.5, respectively, for the time and frequency domain solvers.

In Fig. 2.6, the power flow distribution is close to symmetric at the resonance frequency, 600 THz. As shown in Fig. 2.7, the bulk currents at that resonance frequency are mainly in the silver shell and are flowing from the North pole to the South pole of the CNP, i.e., as J_θ currents proportional to sinθ. These currents produce the fundamental TM dipole mode that this structure radiates.

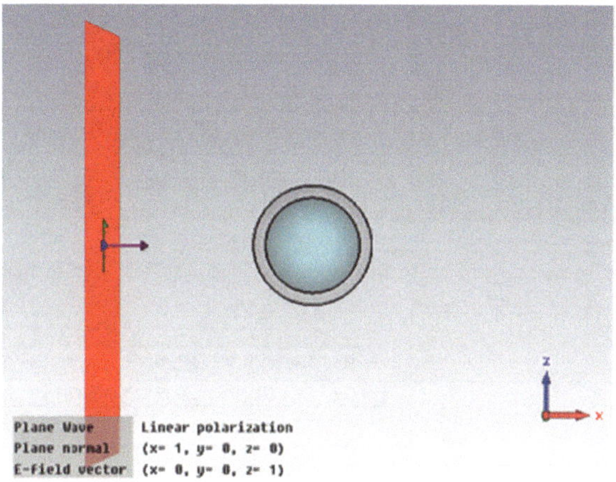

Fig. 2.3 CST CNP model with plane wave source. Figure reproduced with permission from Ref. [6], © 2011 IEEE

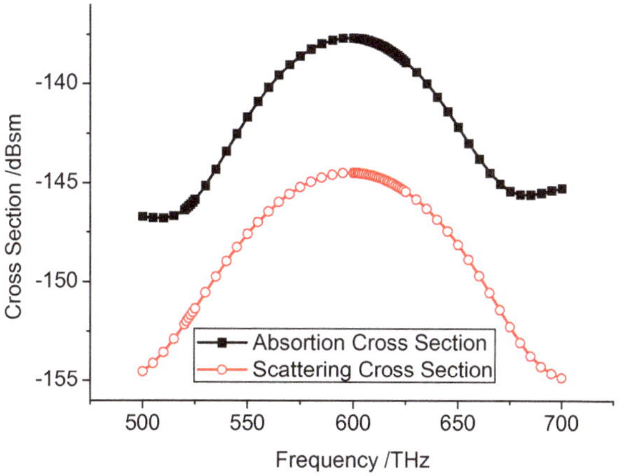

Fig. 2.4 Results from the CST frequency domain solver: absorption cross section and scattering cross section. Figure reproduced with permission from Ref. [2], © 2011 IEEE

2.1.5 Active Open Cylindrical CNP Excited by a Plane Wave

The nano particle considered here is the open cylindrical coated nano-particle, as shown in Fig. 2.8. It consists of either a passive or active cylindrical silica core that is coated on its sides and top with silver, i.e., the silver coating layer forms a cup and is entirely filled with the silica medium. The core is either lossless SiO_2 (passive CNP) or SiO_2 impregnated with gain material (AOC-CNP). This CNP rests its open face on a finite silica slab, the axis of the cylinder being along the normal to the large face of the slab. The coordinate system is shown in Fig. 2.8 b–d. The substrate is taken to be the lossless SiO_2; it is a rectangular slab with dimensions: $2a \times 2a \times dd1$. This would correspond to a silver coating of a simple or gain impregnated silica cylinder grown on a silica slab. We assume that the structure has the fixed height: $L = 31.5$ nm; silver thickness: $dd = 6.0$ nm; and substrate

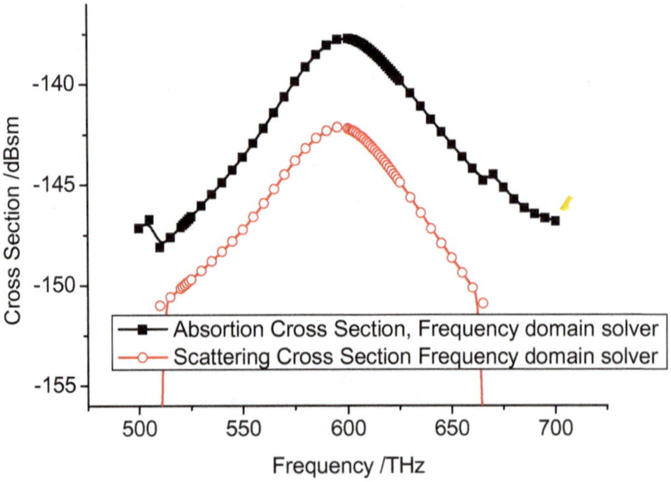

Fig. 2.5 Results from the CST frequency domain solver: absorption cross section and scattering cross section. Figure reproduced with permission from Ref. [2], © 2011 IEEE

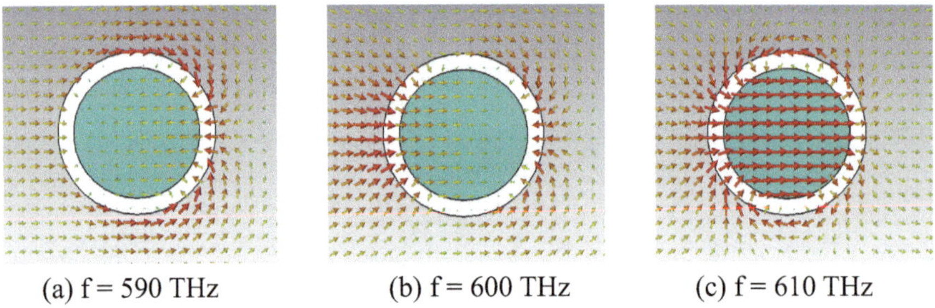

Fig. 2.6 Passive CNP power flow predicted by the frequency domain solver: **a** 590 THz; **b** 600 THz. Figure reproduced with permission from Ref. [2], © 2011 IEEE; **c** 610 THz

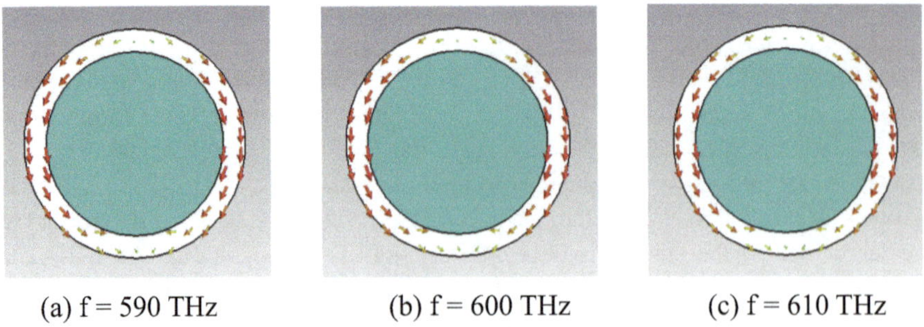

Fig. 2.7 Passive CNP current density predicted by the frequency domain solver: **a** 590 THz; **b** 600 THz. Figure reproduced with permission from Ref. [2], © 2011 IEEE; **c** 610 THz

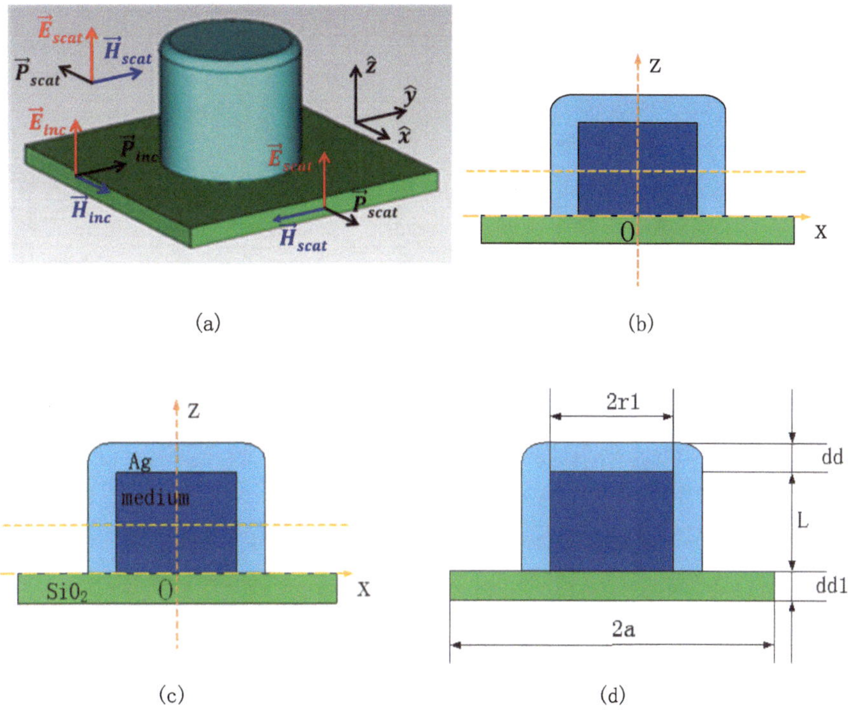

Fig. 2.8 Open cylindrical coated nano particle antenna model: **a** 3D perspective model; **b** y cut plane; **c** material components; **d** dimension and parameters. Figure reproduced with permission from Ref. [6], © 2011 IEEE

thickness: *dd1* = *dd* = 6.0 nm. The silver thickness choice corresponds to the cases considered previously [see, e.g., 3, 8, 9] for the active spherical CNP. The SiO_2 substrate is taken to be 90 nm × 90 nm × 6 nm. The silver cup forms a nano-cavity that is filled with either the passive dielectric or the active gain material.

The entire structure is excited by a vertically polarized plane wave, in which the E field vector is parallel to the z-axis and the propagation direction is along the +x-axis. The incident plane wave field is explicitly given as:

$$\vec{E} = \hat{z}E_0 \cos(\omega t - kx) \tag{2.17}$$

The amplitude of the incident plane wave was $E_0 = 1.0$ V/m.

To determine the resonances of the electrically small, AOC-CNP structure, the CST model was run with the radius r_1 of the inner cavity being swept from 20 to 23 nm. This interval was decided based on our anticipation that the resonance for this open structure would occur for a volume somewhat below the (closed) spherical 30 nm outer radius CNP cases treated, for instance, in [3]. The SCS and ACS cuves are shown in Fig. 2.9. It is clear from the results in Fig. 2.9a, the

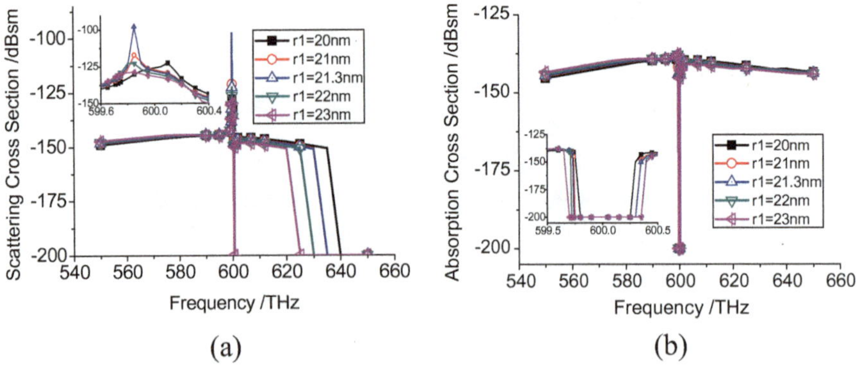

Fig. 2.9 Cross sections with r_1 swept from 20 to 23 nm: **a** Scattering cross section and a zoomed-in view near its maximum values; **b** absorption cross section and a zoomed-in view near its minimum values. Figure reproduced with permission from Ref. [6], © 2011 IEEE

optimum r_1 value is found to be r_1 = 21.3 nm for the AOC-CNP scatterer. And the maximum of the SCS peak: −97.69 dBsm at 599.85 THz is found to this AOC-CNP scatterer, which is more than a 50 dBsm increase over the background. For all simulated values of r_1, the SCS peaks all are actually close to 600 THz in Fig. 2.9a. It is similar that all the ACS curves minimum sinks are close to 600 THz as shown in Fig. 2.9b, and occurs at 599.85 THz. Note that the −200 dBsm floor of the ACS data shown in the subplot of Fig. 2.9b is an artifact of the CST software. Again, the minimum sink is more than 50 dB below the background value. Both the SCS and ACS values clearly show a significant amplification effect of the scattered signal, i.e., the AOC-CNP works as a nano-amplifier around 600 THz, just as the active spherical CNP in [3] did.

In Fig. 2.10, the simulated E-field distributions for the AOC-CNP with r_1 = 21.3 nm at 550.0 THz, 599.85 THz and 650.0 THz are given. The significant enhancement of the electric field in the core of the AOC-CNP at resonance is shown in Fig. 2.10b, and the field vectors form a dipole distribution as expected. Figure 2.11 shows enhancement of the current density in the core at these three frequencies too. In Fig. 2.11b, f = 599.85 THz, t there are very large currents formed in the core at the resonant frequency of the active medium. At other frequency, off the resonance, the current density is large mainly in the shell region. Figure 2.12 are the corresponding magnetic field energy distributions. At the resonance frequency in Fig. 2.12b, the dipole mode of the field is again clearly shown. Note that the energy distribution at resonance is symmetric about the z-axis but is a bit asymmetrical with respect to the middle of the cylinder. The reason is that the open end of the cylinder is not symmetric along the z-axis. Finally, the power flows near the AOC-CNP is shown in Fig. 2.13. In Fig. 2.13b, at resonance, the power is

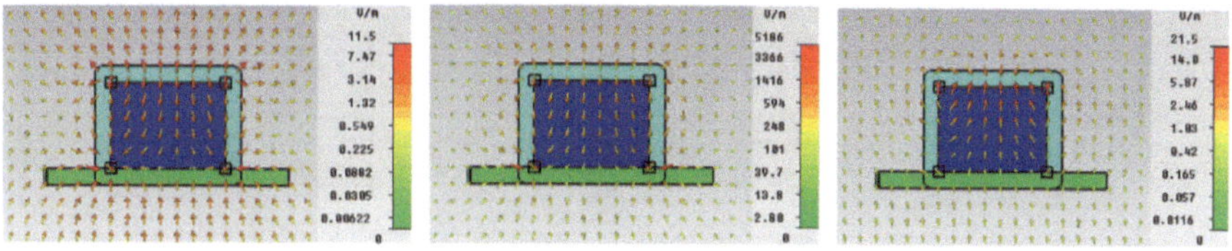

(a) (b) (c)

Fig. 2.10 E-field near the AOC-CNP. **a** f = 550.0 THz; **b** f = 599.85 THz; **c** f = 650.0 THz. Figure reproduced with permission from Ref. [6], © 2011 IEEE

clearly flowing away from the center of the structure. Moreover, it is noted that the power flow within the cylinder experiences a sign change as the frequency varies from below to above its resonance value along with the change in sign of the electric field direction within the core.

2.2 Basic Principle of Spoof Surface Plasmon Polaritons

2.2.1 Dispersion Characteristics of Spoof Surface Plasmon Polaritons Structure

The spoof surface plasmon structure in microwave frequency band is a new type of waveguide transmission line, which realizes the electromagnetic wave fluctuation and motion similar to the surface plasmon wave in optical frequency band through the geometric structure of metal surface. It mainly realizes the surface plasmon effect by etching some unilateral or bilateral groove elements on the transmission line surface. Later, some scholars proposed some complementary structure of SSPPs [10–12]. The dispersion curves of these SSPPs transmission lines are all located on the right side of the light, which indicates that these transmission lines have strong binding ability to electromagnetic waves in the whole frequency band. Among them, the binding capacity of bilateral SSPPs is stronger than that of unilateral SSPPs.

For the traditional unilateral or bilateral SSPPs structure, the depth of the groove has a great influence on the dispersion curve of the traditional unilateral or bilateral SSPPs structure, especially on the electromagnetic wave binding ability. It can be seen from Ref. [13] that the dispersion curve of the traditional bilateral SSPPs structure deviates from the light more and more with the increase of the groove depth *h*, which indicates that the binding ability to electromagnetic waves is enhanced.

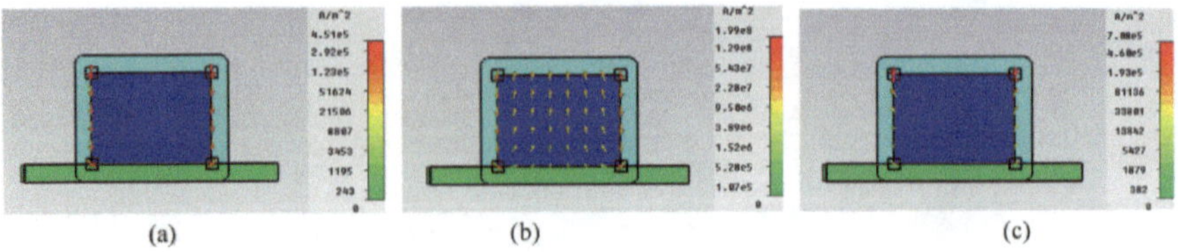

Fig. 2.11 Current density in the AOC-CNP. **a** f = 550.0 THz; **b** f = 599.85 THz; **c** f = 650.0 THz. Figure reproduced with permission from Ref. [6], © 2011 IEEE

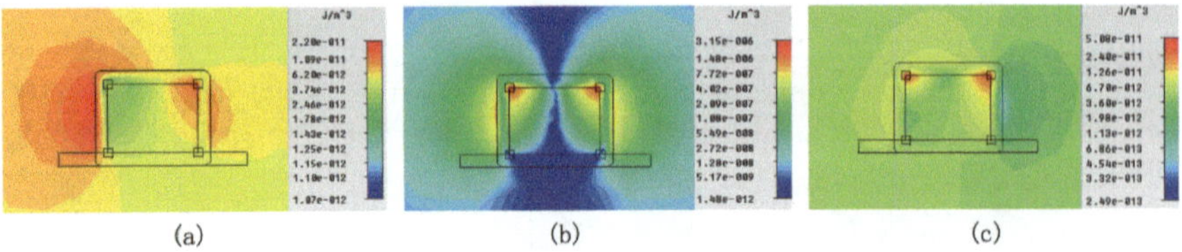

Fig. 2.12 Magnetic field energy density near the AOC-CNP. **a** f = 550.0 THz; **b** f = 599.85 THz; **c** f = 650.0 THz. Figure reproduced with permission from Ref. [6], © 2011 IEEE

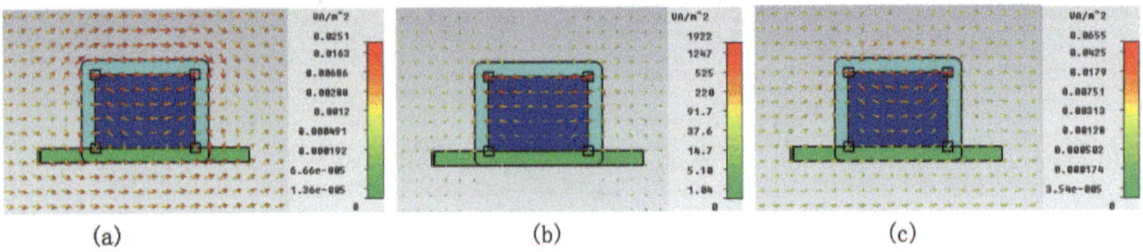

Fig. 2.13 Power flow near the AOC-CNP. **a** f = 550.0 THz; **b** f = 599.85 THz; **c** f = 650.0 THz. Figure reproduced with permission from Ref. [6], © 2011 IEEE

2.2.2 *Basic Characteristics of Spoof Surface Plasmon Polaritons*

2.2.2.1 Strong Binding Properties of Spoof Surface Plasmon Polaritons

The high binding of magnetic field is the characteristic of SSPPs, which is very useful for restraining the mutual coupling between adjacent transmission lines and solving the problem of signal integrity in large-scale circuits [14–16]. By studying the structure of SSPPs, many scholars have designed unilateral or bilateral groove metal units and complementary SSPPs structure, and continuously optimized the transmission performance of SSPPs transmission line [10, 11, 17, 18]. In the direction perpendicular to the propagation direction, the electric field transmitted by the SSPPs transmission waveguide decays exponentially, so it is also used to design the transmission line in the multilayer circuit board to realize the large capacity, high efficiency and low crosstalk transmission of various electromagnetic signals [19].

At the same time, this high binding characteristic of SSPPs also provides a new idea for the design of high gain directional antenna [20].

2.2.2.2 Slow Wave Characteristics of Spoof Surface Plasmon Polaritons

The wave number of the surface plasmon polaritons can be calculated from the dielectric constant of metal and medium on both sides of the interface:

$$k_{\mathrm{SPP}} = \frac{\omega}{c}\sqrt{\frac{\varepsilon_1\varepsilon_2}{\varepsilon_1+\varepsilon_2}} \tag{2.18}$$

Here, ε_1 is the dielectric constant of the metal, ε_2 is the dielectric constant of the medium. The real part of the dielectric constant of metal is generally negative, so the wave number of SPP is always larger than that of light wave.

The dispersion curve of SSPPs is similar to that of SPP. The dispersion curve of SSPPs deviates from light wave, so the wave number of electromagnetic wave transmitted in SSPPs is larger. Generally speaking, k_x the wave number of SSPPs in the propagation direction has the following relationship with k_0 the wave number in the air free space:

$$\alpha_{\mathrm{T}} = \sqrt{k_x^2 - k_0^2} \tag{2.19}$$

Here, α_{T} is the decay constant in the transverse direction (z and y direction). Obviously, with the increase of the wave number in the propagation direction, the binding ability of SSPPs structure to electromagnetic waves will be stronger; on the contrary, if the wave number in the propagation direction is reduced, the binding ability of SSPPs structure to electromagnetic waves will also be weakened.

2.2.2.3 Excitation Method to SSPPs

Due to the great difference of dielectric constant between metal and medium in microwave frequency band, the wave number of SSPPs is much larger than that of traditional guided wave, so it is difficult to directly excite with free space wave or traditional guided wave structure. There are several common incentive methods.

First, the [20–22] SSPPs structure with coaxial inner conductor coupling is a more convenient feeding mode. In Refs. [21, 22], it is proposed to use the coaxial line to excite the SSPPs structure, but the conversion efficiency is low, so it is not suitable for engineering application. Reference [22] places the metal feeder in the middle of the symmetrical SSPPs waveguide to feed the SSPPs waveguide. In Ref. [21], the SSPPs waveguide is improved by using a tapered planar groove structure at the coupling end. This is because reducing the depth of the groove will match the number of electromagnetic waves in the space, which can reduce the reflection of electromagnetic waves. At the same time, by optimizing and adjusting the distance between the feeder and the SSPPs transmission line, the electromagnetic waves can be efficiently coupled into theSSPPps structure.

Secondly, the waveguide fed SSPPs can be realized by introducing the gradient structure. For example, the guided wave in coplanar waveguide (CPW) can be transformed into SSPPs wave [18]. This method has high conversion efficiency and is widely used.

For the complementary [11] or slot SSPPs [12], the waveguide can also be used to feed the SSPPs, and the gap depth adopts gradient excitation SSPPs mode.

2.3 Transmission Mode of Spoof Surface Plasmon Polaritons Structure

SSPPs waveguide can transmit multiple modes, which can be divided into odd mode and even mode according to the symmetrical characteristics of surface current distribution. And it can also be divided into fundamental mode and higher-order mode according to the passband. Transmission modes have different dispersion curves, because different modes have different binding ability to electromagnetic waves. The main propagation modes of SSPPs can be determined by the current distribution or electric field distribution on the surface of SSPPs.

2.3.1 Even Mode of Spoof Surface Plasmon Polaritons

For the traditional symmetrical SSPPs waveguide structure, the electric field near the waveguide structure is completely symmetrical. In this case, the even mode is the main propagation mode in the waveguide. The leaky wave antenna can be realized by modulating on the basis of even mode SSPPs waveguide [13, 23, 24]. Moreover, the radiation beam of this kind of leaky wave antenna is often symmetrical. Reference [13] also proposed that asymmetric structure can also be used to modulate the symmetric SSPPs structure, which can effectively improve the radiation efficiency, but the modulation is essentially based on even mode transmission.

2.3.2 Odd Mode of Spoof Surface Plasmon Polaritons

The odd mode transmission mode of SSPPs has also been studied by many people. In Ref. [10], it is proposed that the odd mode mode can be excited by the combination of asymmetric waveguide and slot structure. The wave number in the odd mode can be adjusted by the shape of complementary structure. The experimental results show that the odd mode transmission mode of SSPPs can effectively restrain the field bound electromagnetic waves. In Ref. [20], a narrow-band SSPPs endfire antenna is realized by using the monopole coupled SSPPs structure. At this time, the odd mode propagation is mainly in the SSPPs waveguide structure.

2.3.3 High-Order Mode of Spoof Surface Plasmon Polaritons

In addition, the waveguide structure based on SSPPs also supports various higher-order modes [25, 26]. Compared with the fundamental mode, the higher-order mode mainly appears at a higher frequency, and there are many standing wave currents in the SSPPs arm.

2.4 Research Methods of Spoof Surface Plasmon Polaritons

2.4.1 Simulation of Dispersion Curve

The design and Simulation in this book mainly uses the commercial electromagnetic field simulation software CST Microwave Studio (CST MWS), which is a 3D full wave electromagnetic field simulation tool. The dispersion curves of SSPPs can be calculated by using the eigenmode solver in CST MWS. Different from other solvers, the eigenmode solver can have no excitation port, but it can not add open boundary while setting periodic boundary conditions. So, it is necessary to add air layer outside the structure to approximate the open boundary. The boundary conditions of dispersion curve simulation in this paper are shown in Fig. 2.14. In the figure, green is the ideal electric boundary condition, and orange is the periodic boundary condition.

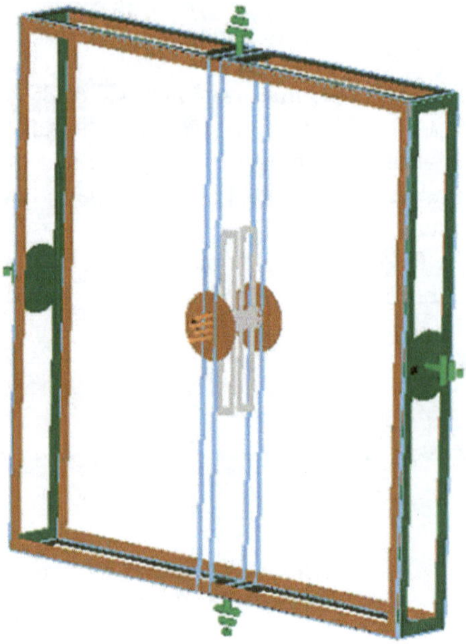

Fig. 2.14 Boundary setting for the cell of SSPPs in CST

2.4.2 *Radar Cross Section Simulation and Calculation*

Radar cross section (RCS) is a measure of the scattering power of the target in a given direction under the illumination of plane wave. For the general scatterer, its own resonance frequency has a high RCS value, so we irradiate the structure to be studied in CST MWS with plane wave to study its scattering characteristics, and obtain the resonance frequency of the structure by looking for the peak value of its RCS curve. The resonant point of the eigenmode of SPP or SSPPs can also be observed by the RCS of the plane wave. Figure 2.15 shows a simulation model of

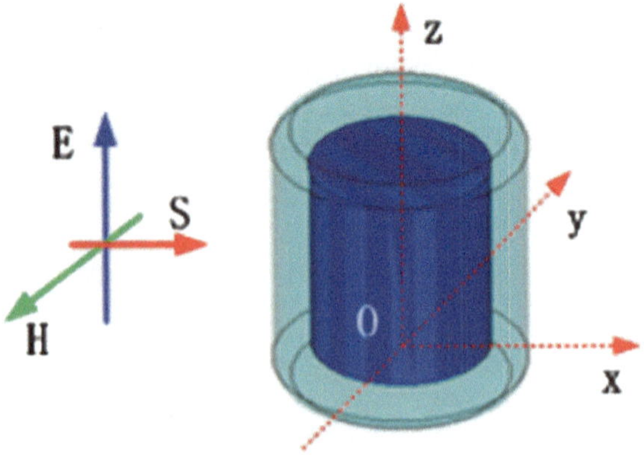

Fig. 2.15 The structure under test and plane wave

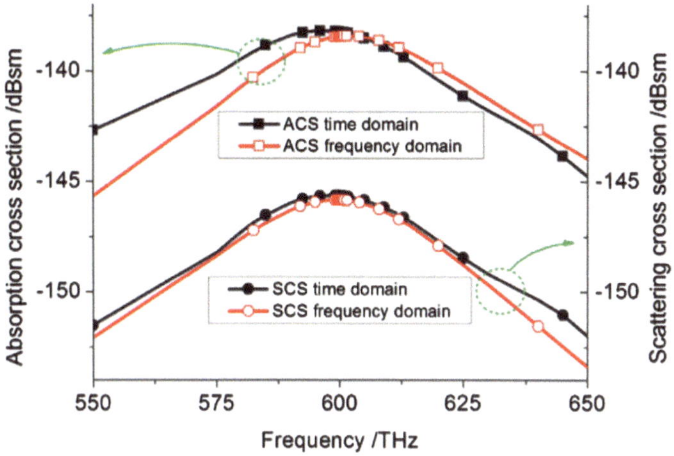

Fig. 2.16 RCS curves of the structure in Fig. 2.14

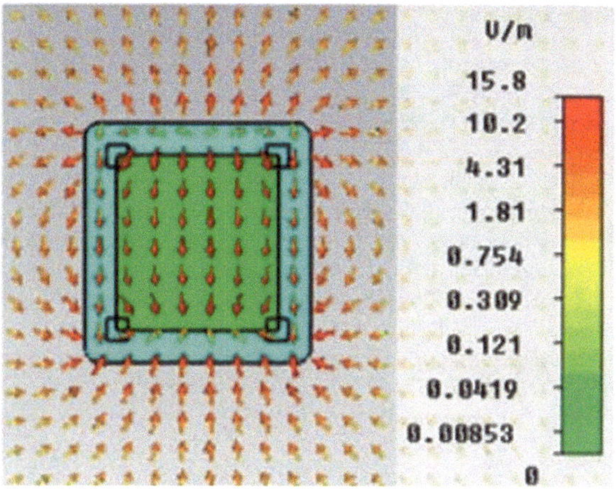

Fig. 2.17 Electric field distribution of the structure in Fig. 2.14

the structure to be tested. The plane wave irradiates from the left side, and the RCS curve is shown in Fig. 2.16. It can be seen that the RCS curve of the structure has a peak at 600 THz, and then the electric field distribution of the structure at 600 THz can be observed, as shown in Fig. 2.17. Obviously, the structure works in TM01 mode at this time.

2.5 Summary of This Chapter

The principle of Surface plasmon polaritons (SPPs) and spoof surface plasmon polaritons (SSPPs) are introduced. Especially the characteristics of the SSPPs structure and the study methods are presented too. This chapter is the basic to the following chapters.

References

1. D. Wang et al., in *A Surface Plasmon Polariton Inspired Patch Antenna.* 2017 IEEE International Symposium on Antennas and Propagation & USNC/URSI National Radio Science Meeting, San Diego, CA, USA, 2017, pp. 2499–2500
2. J. Geng, R.W. Ziolkowski, S. Campbell, R. Jin, X. Liang, in *Studies of Nanometer Antennas Incorporating Gain Material Using CST.* Proceedings of IEEE AP-S International Symposium, Spokane, WA, July 2011, pp. 1624–1627, Paper 308.10
3. J.A. Gordon, R.W. Ziolkowski, The design and simulated performance of a coated nano-particle laser. Opt. Exp. **15**(20), 2622–2653 (2007)
4. P.B. Johnson, R.W. Christy, Optical constants of the noble metals. Phys. Rev. B, Condens. Matter Mater. Phys. **6**(12), 4370–4379 (1972)
5. N.W. Ashcroft, N.D. Mermin, *Solid State Physics* (Holt, Rinehart, Winston, New York, 1976)
6. J. Geng, R.W. Ziolkowski, R. Jin, X. Liang, Numerical study of the near-field and far-field properties of active open cylindrical coated nanoparticle antennas. IEEE Photon. J. **3**(6), 1092–1110 (2011)
7. B.E.A. Saleh, M.C. Teich, *Fundamentals of Photonics* (Wiley, New York, 1991)
8. S. Arslanagi, R.W. Ziolkowski, Active coated nano-particle excited by an arbitrarily located electric Hertzian dipole—resonance and transparency effects. J. Opt. **12**(2), 024014 (2010)
9. S. Arslanagi, R.W. Ziolkowski, Active coated nanoparticles: impact of plasmonic material choice. Appl. Phys. A **103**(3), 795–798 (2011)
10. J.Y. Yin, J. Ren, H.C. Zhang, B.C. Pan, T.J. Cui, Broadband frequency-selective spoof surface plasmon polaritons on ultrathin metallic structure. Sci. Rep. **5** (2015)
11. X. Gao, L. Zhou, T.J. Cui, Odd-mode surface plasmon polaritons supported by complementary plasmonic metamaterial. Sci. Rep. (2015)
12. Y. Zhou, Spoof localized surface plasmons on ultrathin textured MIM ring resonator with enhanced resonances. Sci. Rep. (2015)
13. S.K. Gu, H.F. Ma, B.G. Cai, T.J. Cui, Continuous leaky-wave scanning using periodically modulated spoof plasmonic waveguide. Sci. Rep. **6**, 29600 (2016)
14. J.J Wu et al., Differential microstrip lines with reduced crosstalk and common mode effect based on spoof surface plasmon polaritons. Opt. Express **22**(22), 26777–26787 (2014)
15. H. Zhang et al., *Breaking the Challenge of Signal Integrity Using Time-Domain Spoof Surface Plasmon Polaritons* (2015)
16. P. Cao et al., Multi-layer topological transmissions of spoof surface plasmon polaritons. Sci. Rep. **6**, 22702 (2016)
17. L. Liangliang et al., Dual-band trapping of spoof surface plasmon polaritons and negative group velocity realization through microstrip line with gradient holes. Appl. Phys. Lett. **107** (20), 847 (2015)
18. H. Ma et al., Broadband and high-efficiency conversion from guided waves to spoof surface plasmon polaritons. Laser Photonics Rev. **8**(1), 146–151 (2014)
19. P. Cao et al. Multilayer transmissions of spoof surface plasmon polaritons for multifunctional applications. Adv. Mater. Technol. **2** (2017)

20. Y.J. Han et al., 360° scanning multi-beam antenna based on spoof surface plasmon polaritons. Acta Physica Sinica **65**(14) (2016)
21. X. Shen et al., Conformal surface plasmons propagating on ultrathin and flexible films. Proc. Natl. Acad. Sci. U.S.A. **110**(1), 40–45 (2013)
22. Y. Zhou et al., Bidirectional bending splitter of designer surface plasmons. Appl. Phys. Lett. **99**(11) (2011)
23. Q. Zhang, Q. Zhang, Y. Chen, Spoof surface plasmon polariton leaky-wave antennas using periodically loaded patches above pec and amc ground planes. IEEE Antennas Wireless Propag. Lett. **16**, 3014–3017 (2017)
24. D. Liao, Y. Zhang, H.G. Wang, Wide-angle frequency-controlled beam scanning antenna fed by standing wave based on the cut-off characteristics of spoof surface plasmon polaritons. IEEE Ant. Wirel. Propag. Lett.
25. X. Liu et al., High-order modes of spoof surface plasmonic wave transmission on thin metal film structure. Opt. Express **21**(25), 31155–31165 (2013)
26. L. Zhen et al., High-order localized spoof surface plasmon resonances and experimental verifications. Sci. Rep. **5**, 9590 (2015)

Chapter 3
Multipole Antenna Based on Spoof Surface Plasmon Polaritons Structure

Junping Geng, Han Zhou, and Jingzheng Lu

3.1 Introduction

Since the 1990s, SPP theory has been a hot topic. Untill 2014, the researches on SPP are mainly focused on the application of scatters in optical band, and which is mainly applied in the optoelectronic devices. Paloma [1] explores the spoof magnetic localized surface plasmons with cylindrical structures. Zhou [2] found that based on ultrathin corrugated metal–insulator-metal (MIM) ring resonators, the spoof localized surface plasmon (LSPs) modes can be enhanced. Pendry [3] proposed the concept of surface plasmons with structured surfaces. Since 2014, Xu [4] firstly introduced SPP theory to excite the microwave band radiation modes. Although these studies further improve the SPP theory, but no one applies the theory to the antenna.

In this section, we applied SPP theory to combine work bands of the antenna in different modes, and try to achieve the application of mobile communication band. We firstly study the radiation performances of a SSPPs scatter, and found its main reaonant frequency bands of different SSPPs modes. Then extend the SSPPs structure with feeding part to a low profile SPP antenna, which works at 1.47–2.3 GHz in its master mode. To validate the design, the antenna was simulated and optimized. The simulated results show that the maximum realized gain is 3.7 dB. Later, the improved multipole antenna with different SSPPs modes is proposed. These SSPPs antenna show very good potential as mobile communication antenna.

J. Geng (✉) · H. Zhou · J. Lu
Department of Electronic Engineering, Shanghai Jiao Tong University,
Shanghai 200240, China
e-mail: gengjunp@sjtu.edu.cn

H. Zhou
e-mail: sjtuzhh1993@sjtu.edu.cn

J. Geng et al., *Spoof Surface Plasmon Polaritons Antenna*, Modern Antenna,
https://doi.org/10.1007/978-981-16-4721-5_3

3.2 Multipole Antenna Based on Radial Spoof Surface Plasmon Polaritons Structure

3.2.1 2-D SPP Scatter with Plane Wave in Microwave Band

In the [5], the CNP with surface plasmonic properties has been reported, which can work in a dipole mode at resonant frequency.

Here, according to the [6, 7], the 2-D SPP scatter is constructed as shown in Fig. 3.1, which just like cut the film of the center of the sphere SPP. The substrate is Arlon AD255 cylinder with thickness th = 1 mm and radius r_1 = 30 mm. Above the substrate is the PEC film with thickness 0.1 mm, and the center is a small circle patch with radius r_2 = 5 mm. There are 36 pieces of tapered arms around the circle, and the length of the arm is ld = 25 mm, and every arm occupied 5°.

Here, we radiate the 2-D scatter with plane wave, which can be divided into several kinds.

LP1: linear polarization plane wave with Y direction E field propagated along the X axis

$$\vec{E}_3 = \vec{a}_y E_0 \cos(\omega t - kx) \tag{3.1}$$

LP2: linear polarization plane wave with X direction E field propagated along the Z axis

$$\vec{E}_1 = \vec{a}_x E_0 \cos(\omega t - kz) \tag{3.2}$$

CP3: circular polarization plane wave propagated along Z axis

$$\vec{E}_2 = (\vec{a}_x + j\vec{a}_y) E_0 \cos(\omega t - kz) \tag{3.3}$$

Assume the scatter is circular on the XOY plane, and the scattering can be divided into several cases based on the scatter characters (Fig. 3.2).

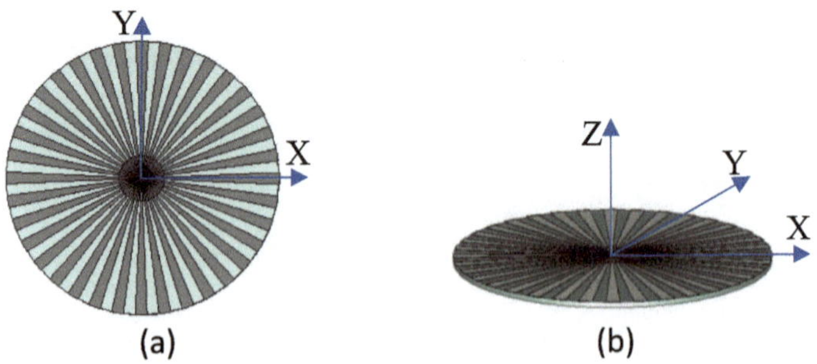

Fig. 3.1 Structure of the 2-D SPP scatter

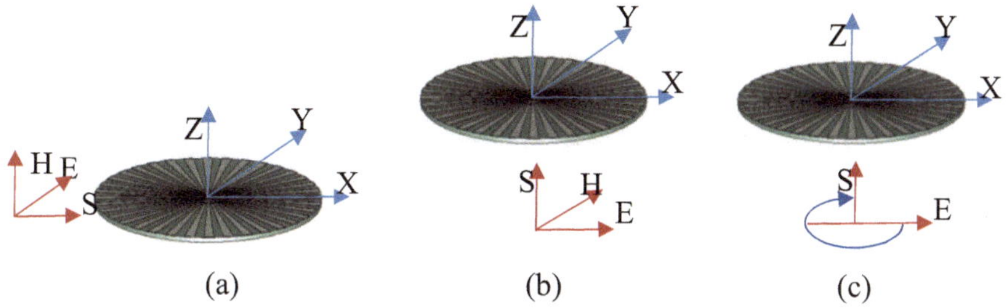

Fig. 3.2 The plane wave radiate the 2-D scatter, **a** case 1; **b** case 2; **c** case 3

Case 1: LP13 radiate the 2-D SPP scatter along horizontal direction (X).

Case 2: LP2 radiate the 2-D SPP scatter along vertical direction (Z).

Case 3: CP2 radiate the 2-D SPP scatter along vertical direction (Z).

In Fig. 3.3, the peaks of the 2D SPP radiated by perpendicular linear polarization plane wave (case 3) or CP plane wave (case 4) or side radiated by linear plane wave (case 5) appear at lower frequency than the conductor patch (case 3, 4) with Arlon255 at same size.

For case 1, the SCS peak of the circular conductor patch with FR4 substrate that radiated by the linear polarization plane wave along Z axis is 8980.12189 mm^2 at f = 2.75 GHz, in which the scattering field works like a E-dipole.

For case 2, the SCS peak of the circular conductor patch with FR4 substrate that grazed by the linear polarization plane wave along X axis is 4423.43966 mm^2 at f = 2.3 GHz, in which the scattering field works like a E-dipole.

For case 3, the SCS peak of the 2D SPP that radiated by the linear polarization plane wave along Z axis is 18188.87 mm^2 at f = 1.61 GHz, in which the scattering field works like a E-dipole at the peak point (Fig. 3.4).

For case 4, the SCS peak of the 2D SPP that radiated by the circular polarization plane wave along Z axis is 18142.412 at f = 1.612 GHz, in which the scattering field works like a E-dipole at the peak point, but it rotated with time, so that it

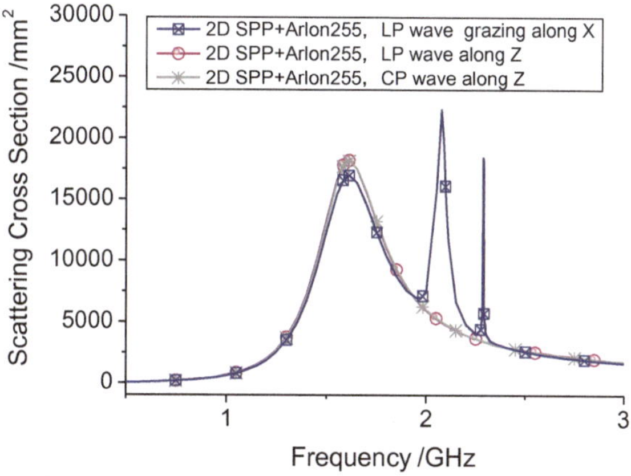

Fig. 3.3 The scattering cross section curves for different cases, there is a clear peak for case 1, case 2 and case 3; for case 1 and 2, there is a very large peak at 1.6 GHz in the SCS curves, especially for case 35, when the linear polarization wave graze along

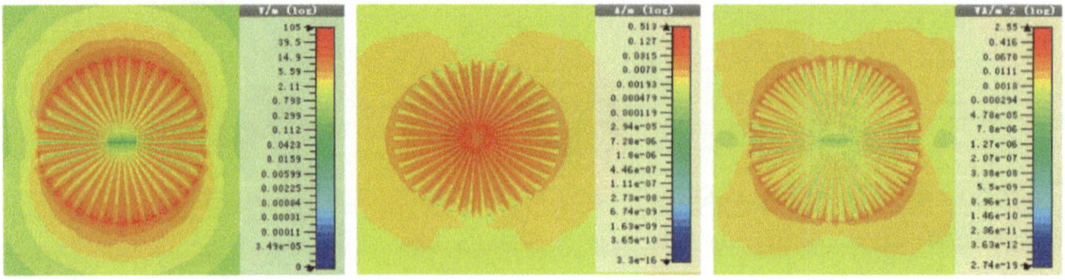

(a) E-field (b) H-field (c) Power flow

Fig. 3.4 Field distribution at f = 1.61 GHz for case 3

shows circular polarization performance, and the power flow is omnidirectional (Fig. 3.5).

For case 5, the linear polarization plane wave just grazed the 2-D scatter. As shown in Fig. 3.6, the whole trend of the SCS curve just like the Gaussian curve with 2 clear stabs, the peaks are 16,957.5211 mm^2 at f = 1.61 GHz, 22,402.4262 mm^2 at f = 2.075 GHz, and 18,503.837 mm^2 at f = 2.285 GHz.

At 1.61 GHz, the scatter like a dipole and the two pole E-filed is excited but a little unsymmetrical. And the far field pattern is still like an apple, which is the E-mode.

At 2.075 GHz, the four-pole E-field is excited, similar as the H-field but bias 45° in azimuth plane, and the far field is four-pole too, which is the 1st M-mode. At 2.285 GHz, the six-pole E-field is excited, and H-field is also six-pole, the far field is six-pole too, this is the 2nd M-mode (Fig. 3.7).

3.2.2 2-D SPP Antenna Fed by Short Stab

As discussed above, the 2D SPP scatter can be excited to strong radiation performance by plane wave. Here, we try to excite the SPP scatter with short stab from the SMA port, and the first SPP antenna structure is shown in Fig. 3.8, and we

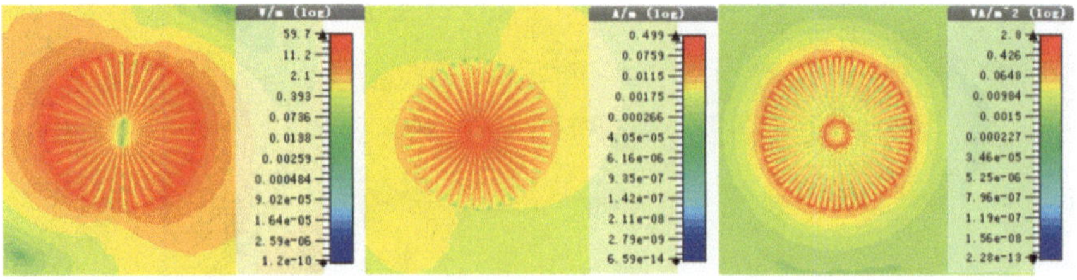

(a) E-field (b) H-field (c) power flow

Fig. 3.5 Field distribution at f = 1.61 GHz for case 4

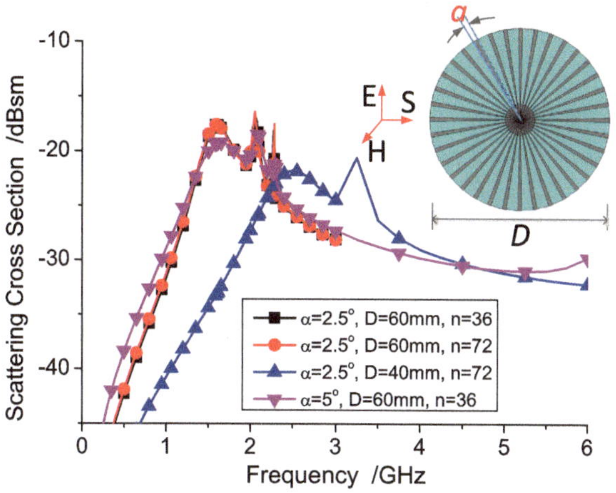

Fig. 3.6 The scattering cross section curves in different cases, α: the angle of every stab; D: the whole diameter of the scatter, and n, the number of the stabs

called it Ant0. The SPP is same size as that in above, the ground is a rectangular with chamfer which just like a part of a circle.

The return loss and realized gain against the frequency are shown in Fig. 3.9, and there are 4 clear work points at f = 1.209 GHz, 1.695 GHz, 2.103 GHz and 2.283 GHz in the S_{11} curve, which are corresponding to the realized gain peaks, whose counterpart EM performance are shown in Fig. 3.10. At f = 1.209 GHz, the E-field is E-dipole mode, which shows 2 poles and symmetrically distributed, which is similar as the case 5 above, mainly is the E-mode. In Fig. 3.10, the E-field mainly is generated between the metal–insulator-metal (MIM) ring and the ground, and just like a dipole.

At f = 1.695 GHz, the E field is dipole mode too, but the H filed biased to the top of the SPP, and E-mode yet. In Fig. 3.10, this E-field is mainly from the metal–insulator-metal (MIM) ring, and corresponding to the E-mode at f = 1.61 GHz in case 5, 6 and 7.

At f = 2.103 GHz, there are four petals in the E field and H field, which means the SPP is excited and work with 4 poles, which is corresponding to M1-mode in case 7 above.

At f = 2.283 GHz, there are six petals in the E field and H field, which means the SPP is excited and work with 6 poles, it is H-mode, the far field is six petals too, which is corresponding to the M2-mode in case 7 above.

3.2.3 2-D SPP Antenna Fed by Stab with Small Open Loop

As shown in Fig. 3.11, the 2-D SPP structure is fed by stab with open loop, the feeding port is 50 Ω, when the stab is excited, the open loop is just like a circular current to radiate the 2-D SPP from back side, which is similar as the circular polarization wave to radiate the 2-D SPP structure in above.

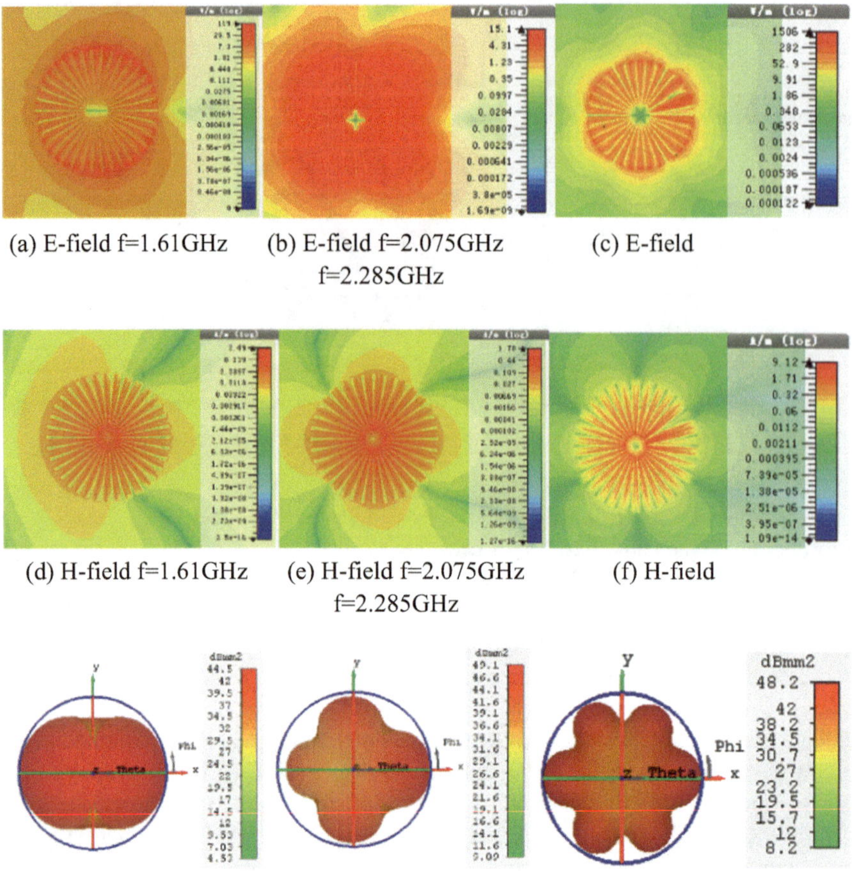

(a) E-field f=1.61GHz (b) E-field f=2.075GHz f=2.285GHz (c) E-field

(d) H-field f=1.61GHz (e) H-field f=2.075GHz f=2.285GHz (f) H-field

(g) far-field f=1.61GHz (h) far-field f=2.075GHz f=2.285GHz (i) far-field

Fig. 3.7 Field distribution for case 5

Fig. 3.8 Structure of linear polarization SPP antenna fed by stab, **a** front side and **b** back side

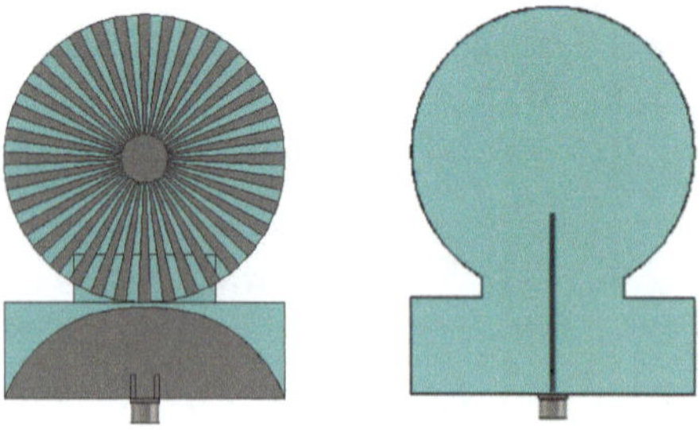

(a) (b)

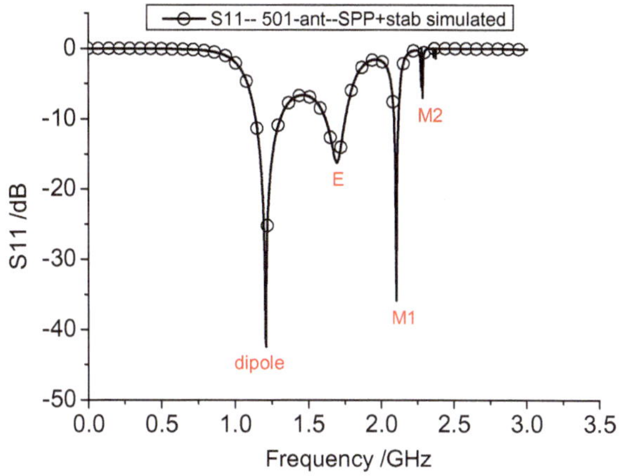

Fig. 3.9 Return loss and realized gain of 2-D SPP antenna fed by stab

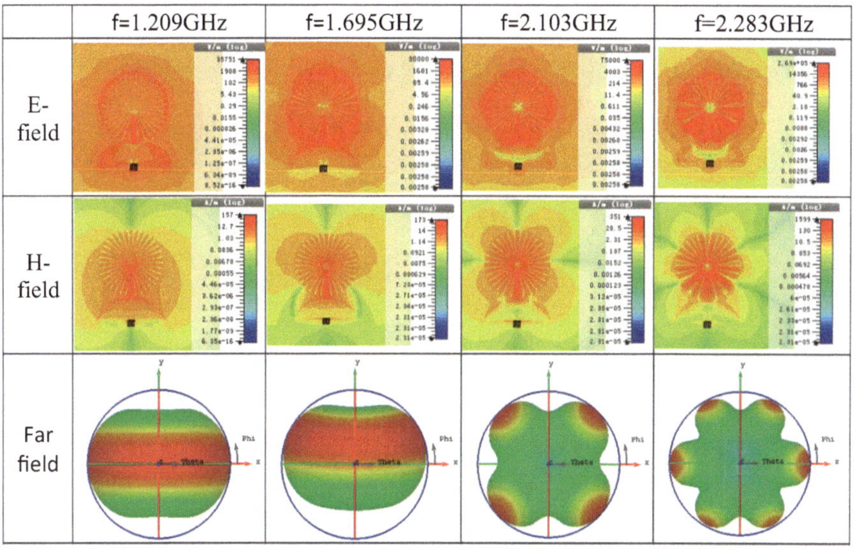

Fig. 3.10 Radiation performance of 2-D SPP antenna fed by short stab

Here, rd = 15 mm, wd = 2 mm, wd1 = 1.2 mm and ld = 20 mm, the 2-D SPP structure and size are same as in above, the ground is a rectangular with chamfer.

The return loss and realized gain against the frequency are shown in Fig. 3.12, and there are 6 clear work points at f = 0.57 GHz, 1.134 GHz, 1.482 GHz, 1.875 GHz, 2.091 GHz and 2.208 GHz in the $S1_1$curve, which are corresponding to the realized gain peaks, whose counterpart EM performance are shown in Fig. 3.13.

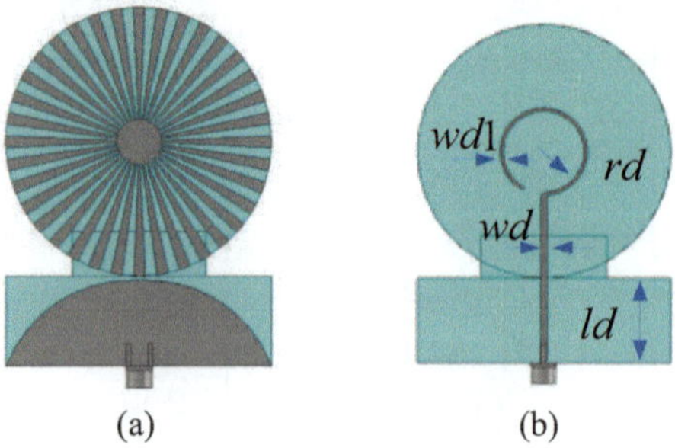

Fig. 3.11 Structure of SPP antenna fed by stab with open loop, **a** front side and **b** back side

At f = 0.57 GHz and 1.134 GHz, the E-field is dipole mode, which shows 2 poles and asymmetrically distributed, and the E-dipole field is mainly radiated out from the slot between the SPP and the ground.

At f = 1.482 GHz, the E field is dipole mode, which is mainly the E mode inspired by SPP, but it rotated with the phase increasing, and it is E-mode yet.

At f = 1.875 GHz, there are 5 petals in the E field and H field, and rotated with phase, which means the SPP is excited and work with 5 poles, which is similar as the H-mode in case 7 above.

At f = 2.091 GHz, there are 7 petals in the E field and H field and rotated with phase, which means the SPP is excited and work with 7 poles, it is H-mode, and similar as the H-mode in case 7 above.

At f = 2.208 GHz, there are 8 petals in the H field and rotated with phase, which means the SPP is excited and work with 8 poles, it is H-mode, and similar as the H-mode in case 7 above.

3.3 Multipole Antenna Based on the Spiral Spoof Surface Plasmon Polaritons Structure

3.3.1 Spoof Surface Plasmon Polaritons Scatter

As shown in Fig. 3.14a, the multi spiral arm SSPP scatterer model is composed of a circular central patch and several equiangular spiral arms surrounding the central patch. In order to study its scattering characteristics, a plane wave is used to irradiate the scattering structure from the back. Figure 3.14b shows the curve of the simulated scattering cross section (SCS) with frequency. It can be seen that the SCS curve of the theoretical model has three peaks at 1.5, 2.25 and 3.3 GHz, which means that the model has strong resonance and scattering characteristics at these three frequencies [8].

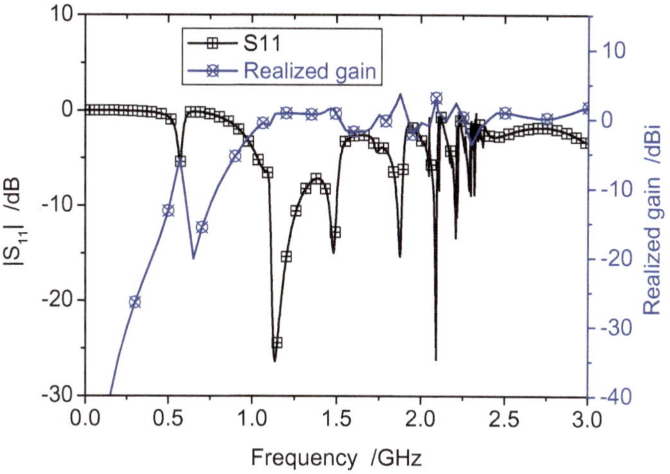

Fig. 3.12 Return loss and realized gain of 2-D SPP antenna fed by stab with open loop

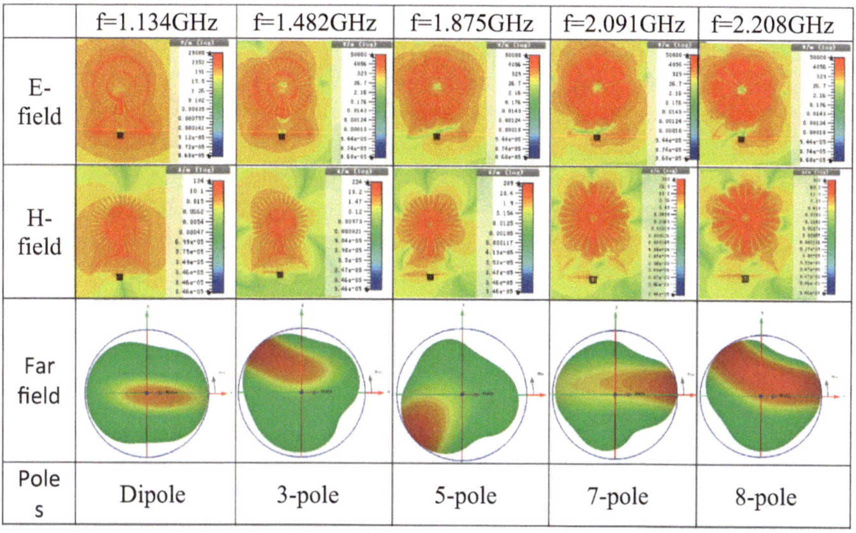

Fig. 3.13 Radiation performance of 2-D SPP antenna fed by stab with open loop

3.3.2 Multipole Antenna Based on Spiral Spoof Surface Plasmon Polaritons Scatter

Based on the above multi spiral arm SSPP scatterer model, a feed ring on the back of the scatter is introduced to construct antenna. In fact, this feeding method is to simulate the plane wave irradiating the scatterer from the back to excite the electromagnetic wave of SSPPs modes. Of course, this feeding form can not completely realize the plane wave irradiation, but we can explore the application of SSPPs scattering model being applied to the specific performance of the antenna. The specific structure of the antenna is shown in Fig. 3.14.

The antenna structure is designed according to the SSPP scatterer model described above, which is shown in Fig. 3.15. The radiation structure of the antenna adopts the multi spiral arms SSPPs scatterer model mentioned above, the ground is a semi-circular metal plate, the dielectric plate of the antenna adopts FR4 dielectric substrate with dielectric constant of 4.4 and loss tangent of 0.016, the radiation structure and the ground of the antenna are etched on the dielectric substrate, the thickness of the metal film is 0.035 mm. There are microstrips with length of W_1 and L_2 on the back of the antenna, and a ring with a radius of R_2. The width of G_2 is used to adjust the input impedance of the antenna to achieve impedance matching. The antenna is energized with a standard 50 Ω SMA connector. The final optimized antenna size is (unit mm) $L_1 = 44$, $L_2 = 17.5$, $D_1 = 3.16$, $W_1 = 22$, $W_2 = 3$, $G_1 = 1.15$, $G_2 = 0.58$, $R_1 = 22$ and $R_2 = 7.7$.

The simulated $|S_{11}|$ of the antennais shown in Fig. 3.16. It can be seen that the reflection coefficients of the antenna at 1.5, 2.1 and 2.9 GHz all have minimum values, and have two resonances with −10 dB impedance bandwidths equal to 1.47–2.28 and 2.7–2.9 GHz. Figure 3.17 shows the three-dimensional patterns of the antenna at 1.5 GHz, 2.1 GHz and 2.9 GHz, respectively.

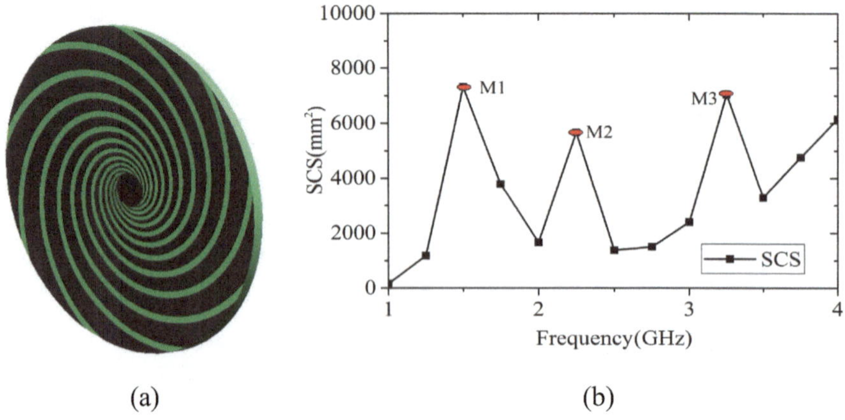

Fig. 3.14 The spiral SSPP scatterer model and far field SCS. **a** Scatter model; **b** SCS curve with peaks at 1.5, 1.85 and 2.22 GHz. Figure reproduced with permission from Ref. [8], © 2017 IEEE

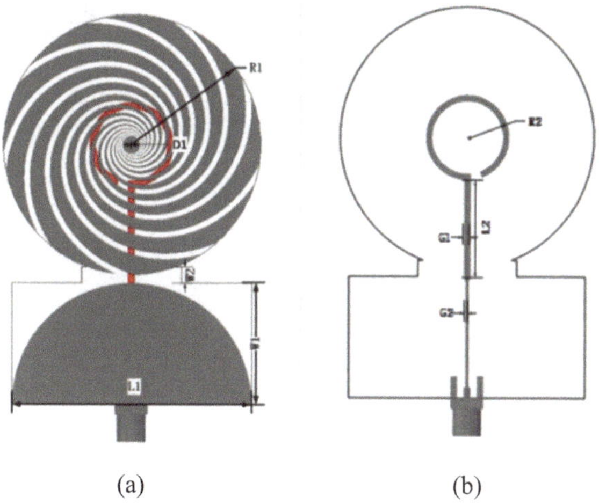

Fig. 3.15 SSPP antenna model. **a** Top view; **b** back view. Figure reproduced with permission from Ref. [8], © 2017 IEEE

From the analysis of the above results, it can be seen that the peak value of the simulated SCS of the SSPPs scatterer with multi spiral arms corresponds to the resonance point of the antenna one by one, which indicates that it is feasible to introduce the SSPPs scatterer model with multi spiral arms into the antenna structure. It can also be seen from the pattern in Fig. 3.17 that when the antenna works in SSPPs modes, different modes will have different radiation patterns, even if the resonance points of two different operating bands are very close. This provides ideas for the design of the following antenna form. According to the multi-mode characteristics of SSPPs antenna, an improved three band antenna to WLAN and Winmax is designed based on SSPPs theory.

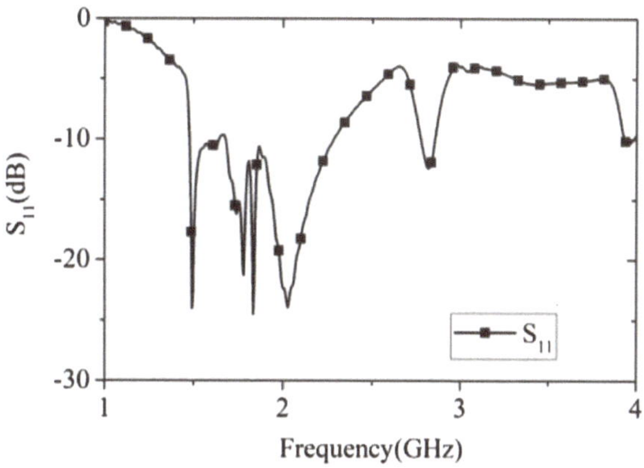

Fig. 3.16 Simulated S parameters of the antenna. Figure reproduced with permission from Ref. [8], © 2017 IEEE

Fig. 3.17 The three dimension radiation pattern of the antenna. **a** 1.5 GHz; **b** 2.1 GHz; **c** 2.9 GHz. Figure reproduced with permission from Ref. [8], © 2017 IEEE

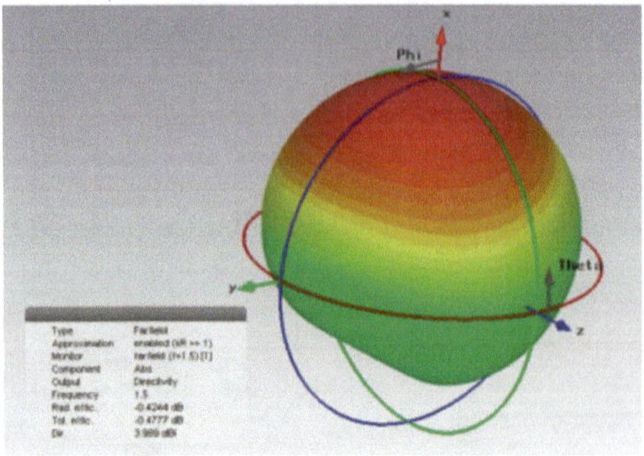

(a)

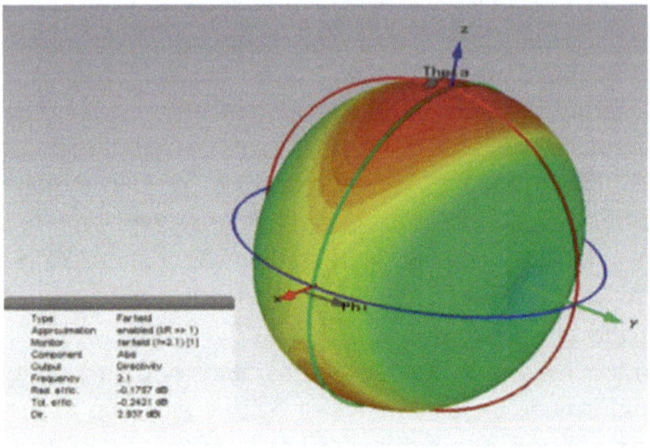

(b)

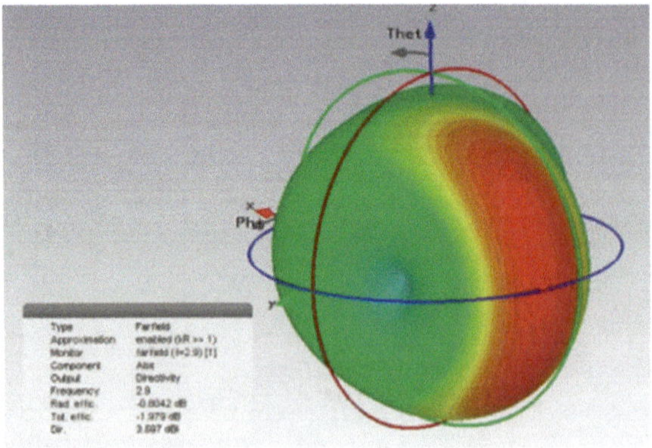

(c)

3.4 Summary of This Chapter

In this chapter, the multipole antennas based SSPPs structure are presented.

Based on the multi radial arm structure, the 2-D surface plasmonic polarition structure is studied, in which the E-dipole and M-dipole modes can be excited at sub-wavelength, and the resonant frequencies move to lower band in microwave band. By introducing the feeding stab, the 2-D SSPPs antenna can work in microwave band with E-dipole mode and several M-dipole modes. In further, the 2-D SPP antenna fed by stab with small open loop is studied, in which more M-dipole modes are excited, and they work in L-band.

Based on the multi spiral arm structure, the 2-D surface plasmonic polarition structure is studied, in which the E-dipole and M-dipole modes can be excited at sub-wavelength, and the resonant frequencies move to lower band in microwave band. By introducing the feeding stab with small open loop, the 2-D SSPPs antenna with spriral arms can work in microwave band with E-dipole mode and several M-dipole modes.

References

1. P.A. Huidobro, E. Moreno, L. Martin-Moreno, F.J. Garcia-Vidal, in *Magnetic Localized Surface Plasmons*. 8th International Congress on Advanced Electromagnetic Materials in Microwaves and Optics, Lyngby, 2014, pp. 340–342
2. Y.J. Zhou, Q.X. Xiao, B.J. Yang, Spoof localized surface plasmons on ultrathin textured MIM ring resonator with enhanced resonances. Sci. Rep. **5**(14819) (2015). https://doi.org/10.1038/srep14819
3. J.B. Pendry, L. Mart-Moreno, F.J. Garcia-Vidal, Mimicking surface plasmons with structured surfaces. Science **305**, 847–848 (2004)
4. J.J. Xu, H.C. Zhang, T.J. Cui, Efficient conversion of surface-plasmon-like modes to spatial radiated modes. Appl. Phys. Lett. **106**, 021102 (2015). https://doi.org/10.1063/1.4905580
5. R.W. Ziolkowski, S. Arslanagic, J. Geng, in *Where High-Frequency Engineering Advances Optics: Active Nano-Particles as Nanoantennas*, by ed. A. Alu, M. Agio. Optical Antennas, Chap. 4 (Cambridge University Press, New York, 2013)
6. P.A. Huidobro, X. Shen, J. Cuerda, E. Moreno, L. Martin-Moreno, F.J. Garcia-Vidal, T.J. Cui, J.B. Pendry, Magnetic localized surface plasmons. Phys. Rev. **X 4**, 021003 (2014)
7. F. Qin, J.J. Xiao, *Progress in Electromagnetic Research Symposium (PIERS)—Unidirectional RF Antenna by Stacked Spoof Surface Plasmon Resonators*, pp. 1680–1680 (2016) (IEEE 2016 Progress in Electromagnetic Research Symposium (PIERS)—Shanghai, China (2016.8.8–2016.8.11)])
8. D. Wang, J. Geng, X. Liang, W. Zhu, C. He, in *A Novel Patch Antenna Based on Surface Plasma Polarization*. IEEE 5th International Symposium on Electromagnetic Compatibility (EMC-Beijing), Beijing, 2017, pp. 1–2

Chapter 4
Endfire Antenna Based on Spoof Surface Plasmon Polaritons

Junping Geng, Kaijie Zhuang, and Silei Yang

Due to the strong binding effect of artificial surface plasmons on electromagnetic field, the early artificial surface plasmons were mainly used in the design of transmission lines. Because the phase of artificial surface plasmons does not match the free space, artificial surface plasmons cannot directly radiate electromagnetic waves to the free space. But this strong binding ability to electromagnetic field can be used to reduce the coupling between antennas [1], so artificial surface plasmon has also been applied in the field of antenna. At present, most of the antennas based on artificial surface plasmons work in the side fire mode [2–4]. A few literatures have reported the end fire antennas based on artificial surface plasmons. On the basis of the existing research, this chapter studies the asymmetric feed SSPPs end fire antenna, and designs a broadband, high efficiency end fire antenna based on artificial surface plasmon.

4.1 Establishment and Simulation of Antenna Model

4.1.1 Antenna Model

The electromagnetic model of the end fire antenna based on SSPPs is shown in Fig. 4.1, which mainly includes two parts: an asymmetric coplanar waveguide structure and two symmetrical unilateral SSPPs metal structures with a narrow slot in the middle. All metals are attached to the FR-4 plate with a thickness of 1 mm and the loss tangent of 0.025. Referring to Ref. [5], both ends of the single side SSPPs metal structure are highly graded. The whole size of the antenna is 210 mm * 44 mm * 1 mm, and the center frequency is 2.6 GHz.

J. Geng (✉) · K. Zhuang · S. Yang
Department of Electronic Engineering, Shanghai Jiao Tong University, Shanghai, China
e-mail: gengjunp@sjtu.edu.cn

J. Geng et al., *Spoof Surface Plasmon Polaritons Antenna*, Modern Antenna,
https://doi.org/10.1007/978-981-16-4721-5_4

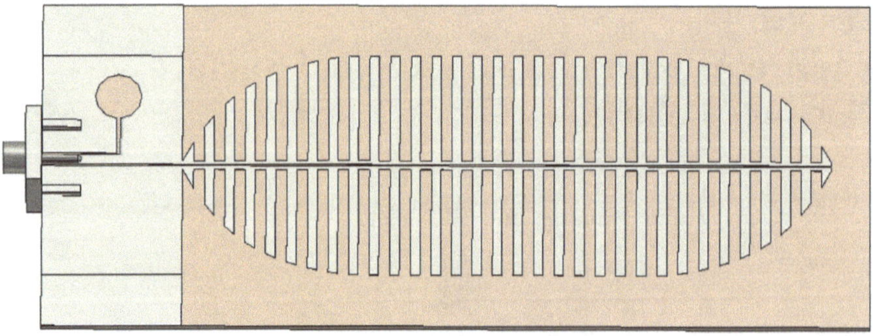

Fig. 4.1 Initial antenna model

As shown in Fig. 4.1, a circular cavity is introduced into the gap on one side of the coplanar waveguide to realize asymmetric feeding. In this way, the current flowing into the two unilateral SSPPs structures is approximately equal amplitude reverse, which is equivalent to the transmission of odd mode SSPPs wave. After the electromagnetic energy enters through the SMA joint, the phase difference between the two sides is adjusted by the circular cavity, and then it is coupled into the SSPPs structure from the gap of the coplanar waveguide, and propagates forward in the form of traveling surface wave. After reaching the end, it radiates to the free space through the tapered structure at the end. SSPPs structure has a strong binding effect on the electromagnetic field, so that the end beam of the antenna will be narrow, and the gain will be relatively high. In addition, due to the ground effect of coplanar waveguide, the back radiation of the antenna will be suppressed and the back lobe will be smaller.

On the other hand, from the mode of SSPPs structure, the traditional symmetrical bilateral SSPPs structure generally adopts the gradual change structure from coaxial transition to symmetrical coplanar waveguide. The symmetrical bilateral SSPPs structure mainly transmits even modes, and there is no end radiation. In the design of Fig. 4.1, the asymmetric coplanar waveguide transition is used to feed the bilateral SSPPs structure. The electric field on the SSPPs structure should be antisymmetric in order to realize odd mode transmission and high gain end fire radiation performance.

4.1.2 Antenna Performance and Analysis

4.1.2.1 S Parameters

The simulated S-parameter results of this antenna are shown in Fig. 4.2. The frequency range of $|S_{11}| < -10$ dB is 2.6–2.7 GHz, and the bandwidth is a little narrow. Compared with the existing SSPPs antenna, the advantage is not obvious.

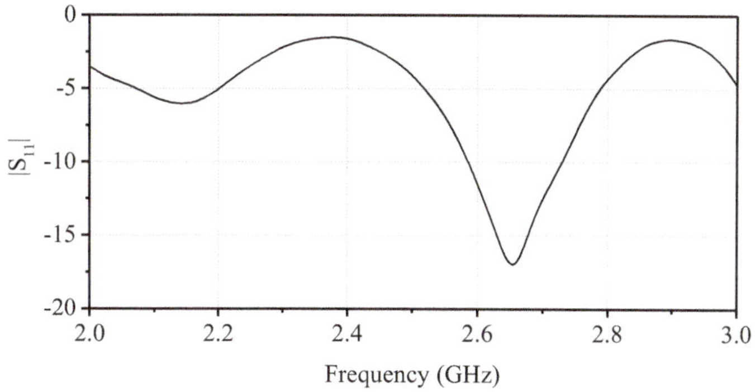

Fig. 4.2 The simulated $|S_{11}|$ of proposed antenna

However, this design method of transforming symmetric feed into asymmetric feed is very valuable, which is also the basis of our subsequent design of high-performance end fire SSPPs antenna.

4.1.2.2 E-field

At $f = 2.65$ GHz, the Z component distribution of the electric field $\mathbf{E}_z$ of the antenna is shown in Fig. 4.3. It can be seen that the electric field $\mathbf{E}_z$ is antisymmetric with respect to the *xoz* plane, and the odd mode SSPPs wave is excited in the SSPPs structure, which is consistent with the working principle of the previous analysis.

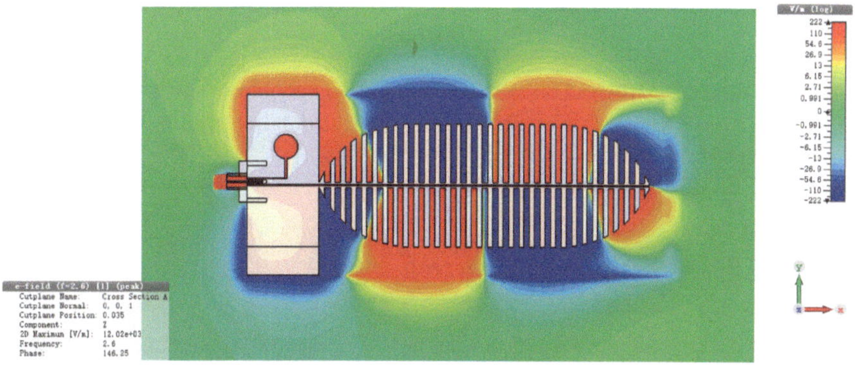

Fig. 4.3 The simulated electric distribution (Z component) of proposed antenna

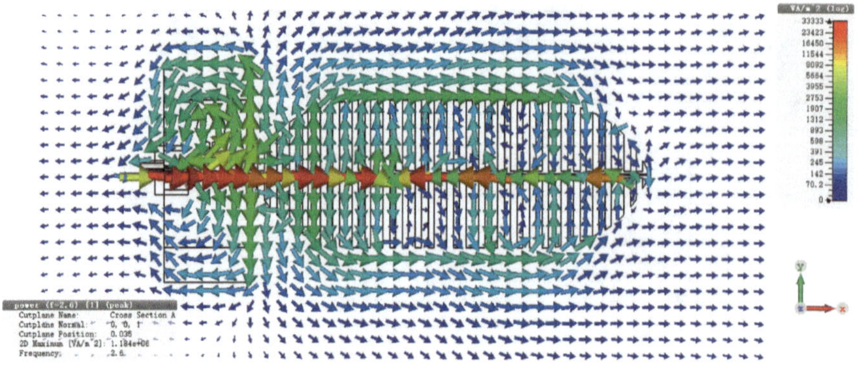

Fig. 4.4 The simulated power flow of proposed antenna

4.1.2.3 Power Flow

The power flow distribution of the antenna in *xoy* plane is shown in Fig. 4.4. Obviously, the energy in the gap is gradually coupled into the SSPPs structure. In addition, a part of the energy is leaked out at the junction of coplanar waveguide and SSPPs structure, but due to the coupling effect of SSPPs, this part of the electromagnetic energy is still radiated to the end. In fact, the main electromagnetic energy is still bound in the SSPPs structure, and radiates to the free space at the end.

4.1.2.4 Far Field

The radiation pattern of the SSPPs antenna is shown in Fig. 4.5, which is obviously end fire. The simulated realized gain of the antenna is 6.96 dB, and the back lobe is very small. The simulation results show that the radiation efficiency is higher than 80%.

Different from the existing SSPP end fire antenna, the propsed SSPPs end fire antenna proposed in this section is fed by asymmetric coplanar waveguide structure, and the end fire effect is achieved skillfully. The simulation results show that this feeding method can realize SSPPs end fire antenna.

4.2 Design of Improved Antenna Model

In order to increase the bandwidth and gain of SSPPs end fire antenna, this section improves the antenna model based on the previous SSPPs antenna model. Especially, the feed part is directly fed by slot coupling. The details will be discussed as follows [6].

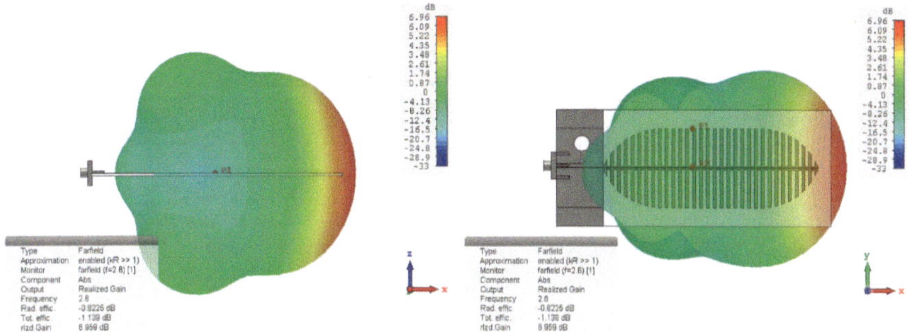

Fig. 4.5 The simulated radiation pattern of proposed antenna

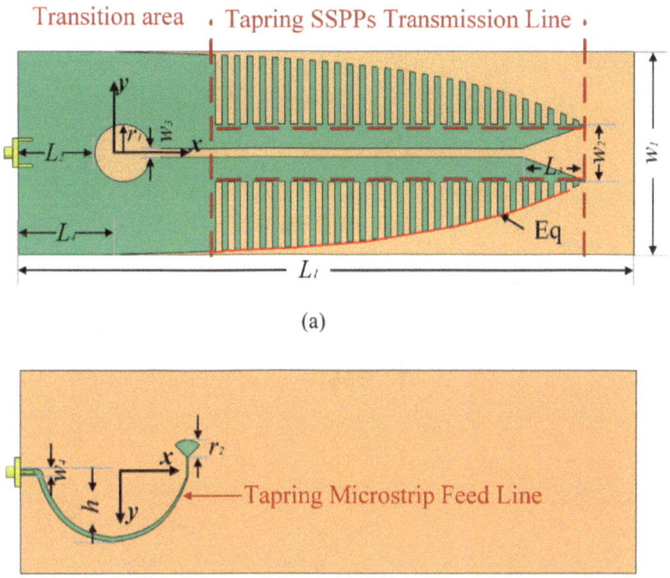

Fig. 4.6 Proposed endfire radiation antenna. **a** Front view; **b** back view of the proposed antenna. Reproduced courtesy of The Electromagnetics Academy

4.2.1 Antenna Structure and Principle

The front view and back view of the optimized antenna model are shown in Fig. 4.6, in which the green part represents metal and the yellow part represents medium. The antenna consists of transition structure and SSPPs structure. The metal part of the antenna is printed on the 0.8 mm thick Arlon AD250 dielectric substrate (the dielectric constant is 2.5 and the loss tangent is 0.0018). The feed part is on the back of the antenna, and the arc-shaped microstrip line with variable width is used to achieve wide impedance matching. A circular cavity and a triangular slot are connected at the end of the slot respectively. The electromagnetic energy of the gap will be gradually coupled into the SSPPs structure. The overall size of the antenna is 210 mm × 70 mm × 0.8 mm.

The Z component distribution of the electric field of the antenna at 2.5 GHz is shown in Fig. 4.7. It can be seen that the odd mode is excited in the SSPPs transmission line. In addition. As the electromagnetic wave propagates to the right end, the depth of the groove becomes shallower, and the period of the electromagnetic wave becomes longer, which means that the wave number becomes smaller, the binding ability of SSPPs structure to the electromagnetic field is gradually weakened, and the leakage of electromagnetic wave is enhanced.

The power flow distribution of the antenna at 2.5 GHz is shown in Fig. 4.8. It can be seen that in the central region, the EM energy is continuously coupled from

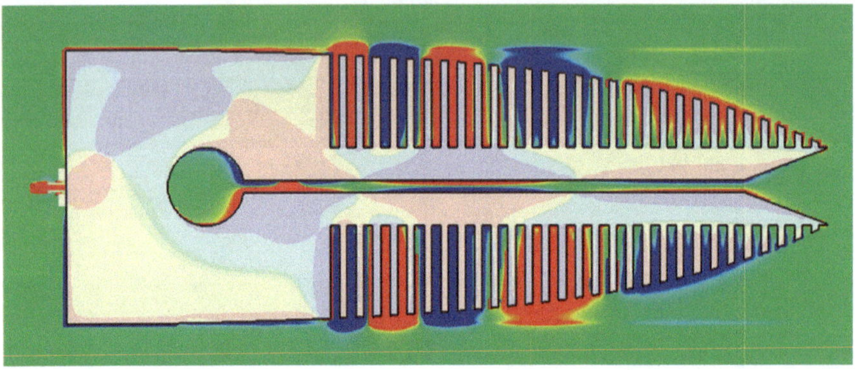

Fig. 4.7 The simulated electric filed distribution of Z component at 2.5 GHz. Reproduced courtesy of The Electromagnetics Academy

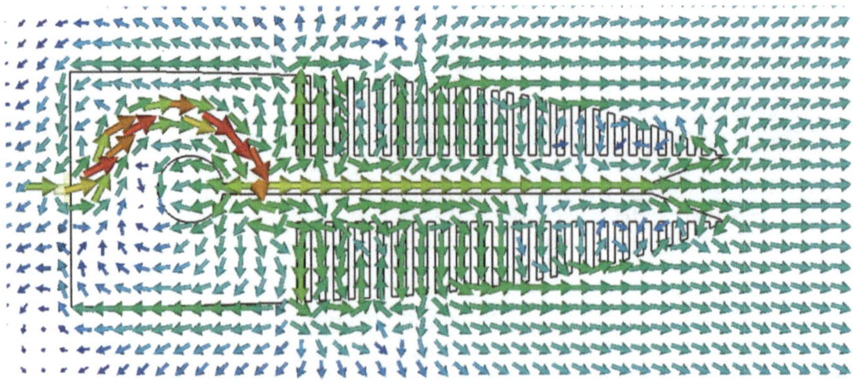

Fig. 4.8 The simulated power flow distribution at 2.5 GHz. Reproduced courtesy of The Electromagnetics Academy

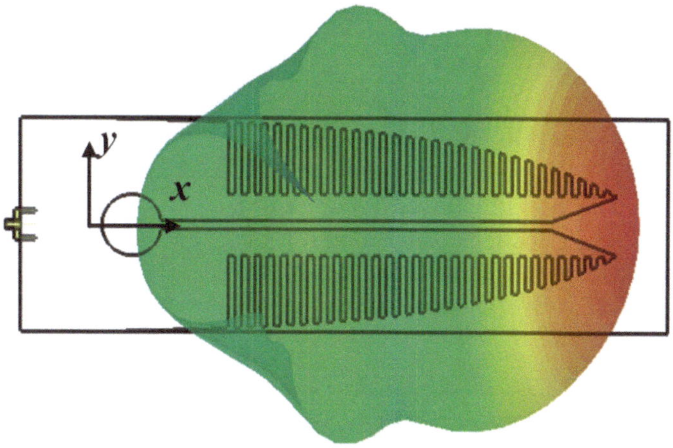

Fig. 4.9 The simulated radiation pattern at 2.5 GHz. Reproduced courtesy of The Electromagnetics Academy

the slot to the SSPPs structure, and the SSPPs structure and the triangular slot together form the radiation part of the antenna. Figure 4.9 shows the far-field pattern of the antenna at 2.5 GHz. It is obvious that the rear lobe of the antenna is very small, and the end radiation energy is mainly concentrated in the main lobe.

4.2.2 *Simulation Verification and Parameter Analysis*

This section mainly discusses the influence of the key parameters of the antenna on the performance of the antenna.

The height h(x) of SSPPs structure changes exponentially with x, and its contour can be expressed by the following curve equation:

$$h(x) = \exp(\ln(0.5W_1 + 1)/(L_1 - L_4) * (x - L_4)) - 1 - 0.5W_1$$
$$(L_4 \leq x \leq L_1) \quad (4.1)$$

4.2.2.1 W_1

Figure 4.10 shows the simulation curve of antenna return loss $|S_{11}|$ with different width W_1. It is obvious that with the increase of W_1, the working frequency band of the antenna gradually moves to lower frequency, and the overall bandwidth increases, but some frequency points deteriorate. At the same time, when

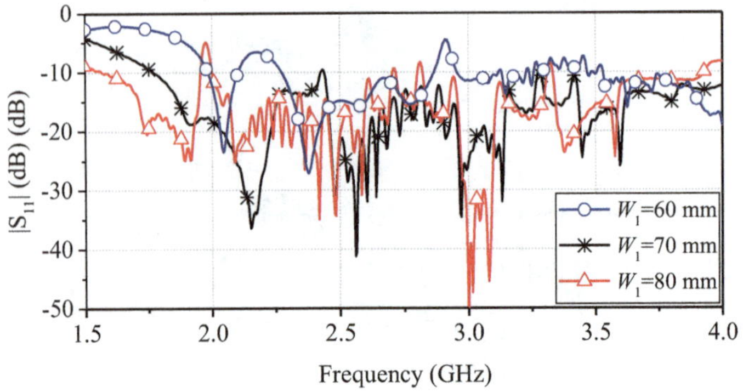

Fig. 4.10 Effect of W_1 on the $|S_{11}|$ performance. Reproduced courtesy of The Electromagnetics Academy

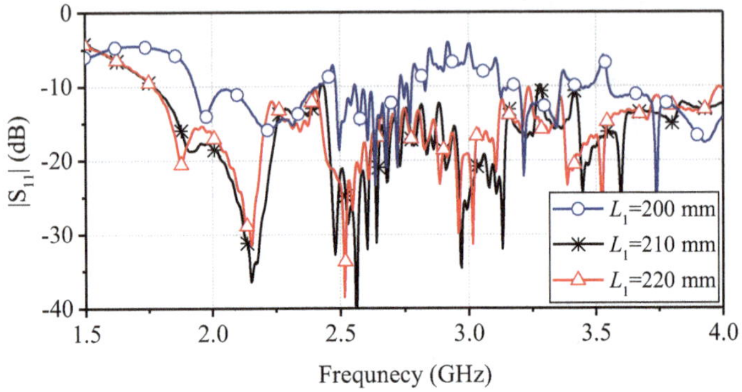

Fig. 4.11 Effect of L_1 on the $|S_{11}|$ performance. Reproduced courtesy of The Electromagnetics Academy

$W_1 = 60$ mm, the return loss $|S_{11}|$ curve deteriorates as a whole. Finally, we choose $W_1 = 70$ mm.

4.2.2.2 L_1

In addition, by changing L_1 for simulation, the simulated curve of antenna return loss $|S_{11}|$ at different widths of L_1 is shown in Fig. 4.11. It can be seen that when L_1 is 210 or 220 mm, the influence of L_1 on antenna impedance bandwidth is not obvious, but when L_1 is reduced to 200 mm, the $|S_{11}|$ of antenna deteriorates obviously. Finally, we choose $L_1 = 210$ mm.

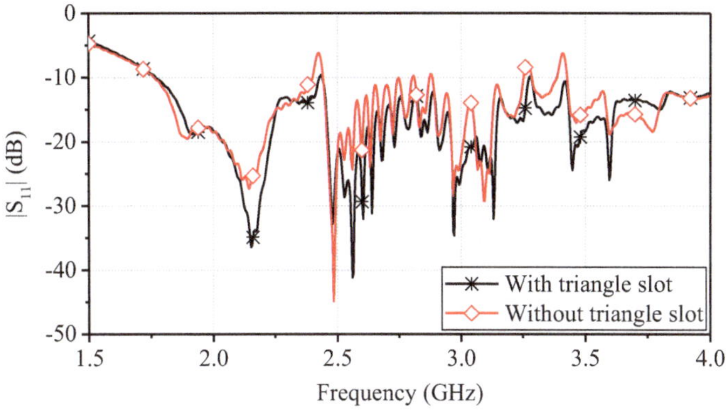

Fig. 4.12 Comparison of $|S_{11}|$ performance with and without the triangle slot at the end. Reproduced courtesy of The Electromagnetics Academy

4.2.2.3 Triangular Slot

The $|S_{11}|$ curve of the antenna is also affected by the introduction of triangular slot or not at the end of the antenna. The specific simulation comparison is shown in Fig. 4.12. It can be seen that the return loss $|S_{11}|$ of the antenna can be significantly improved by introducing a triangular slot. This is because the introduction of a triangular open slot is equivalent to expanding the aperture at the end of the traveling wave waveguide with a slender slot. Through this, the effective radiation of energy can be increased and the reflection can be reduced.

4.2.2.4 Effect of the SSPPs Structure

Here, we mainly discuss the effect of gradual SSPPs structure on antenna performance. Figure 4.13 shows a narrow slot structure without SSPPs structure, which is essentially a slot waveguide. Figure 4.14 shows the power flow distribution of the

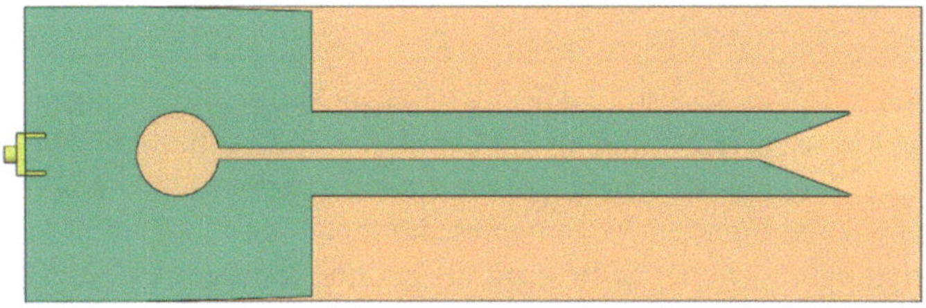

Fig. 4.13 Narrow slot antenna. Reproduced courtesy of The Electromagnetics Academy

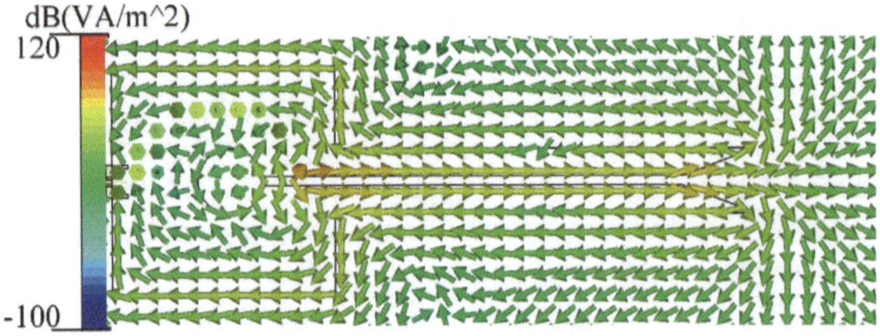

Fig. 4.14 Power flow of the narrow slot antenna at 2.5 GHz. Reproduced courtesy of The Electromagnetics Academy

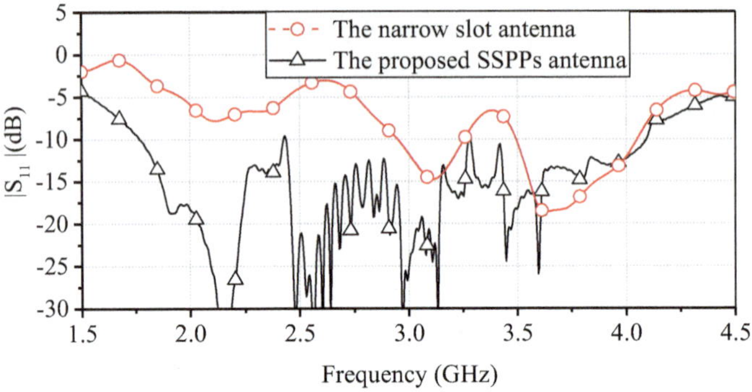

Fig. 4.15 Comparison of the $|S_{11}|$ performance between the narrow slot antenna and the proposed SSPPs antenna. Reproduced courtesy of The Electromagnetics Academy

Table 4.1 The optimized antenna parameters

Parameters	h	L_1	L_2	L_3	L_4	r_1
Values/mm	25	216	27.1	20.6	29	10
Parameters	r_2	W_1	W_2	W_3	W_4	
Values/mm	6	70	3	2.1	19.6	

Reproduced courtesy of The Electromagnetics Academy

slotted waveguide structure at 2.5 GHz. Obviously, the energy confined in the long and narrow gap is divided into two parts at the end of the structure. One part radiates out to form the end beam, and the other part propagates along the negative direction of x-axis. Figure 4.15 compares the return loss $|S_{11}|$ of the narrow slot waveguide structure without SSPPs and the endfire antenna with SSPPs. It can be seen that the working bandwidth of the long and narrow slot waveguide structure is very narrow, and it can only provide directional end beam in the narrow band.

After simulation and optimization, the optimal parameters of the antenna are shown in Table 4.1. The optimized antenna parameters.

(a)

(b)

Fig. 4.16 The fabricated prototype

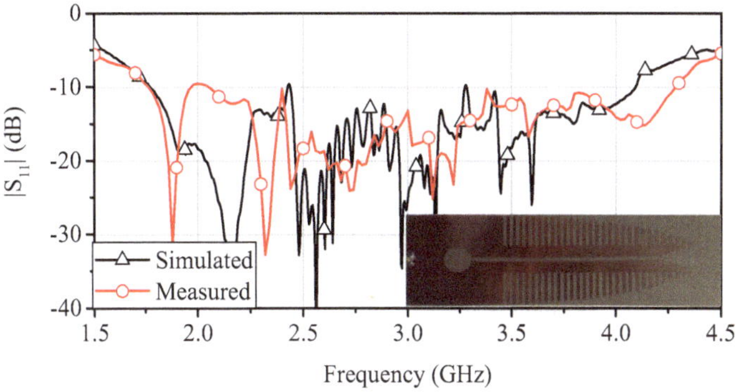

Fig. 4.17 Comparison of the simulated and measured $|S_{11}|$ values of the proposed endfire antenna. Reproduced courtesy of The Electromagnetics Academy

4.2.3 Experimental Result and Discussion

According to the optimized parameters, the antenna prototype is manufactured and its far-field pattern is tested in Satimo anechoic chamber. Figure 4.16 shows the actual antenna processed.

Figure 4.17 shows the simulated and measured antenna return loss $|S_{11}|$ curves. It can be seen that the measured impedance bandwidth is from 1.8 to 4.3 GHz, and

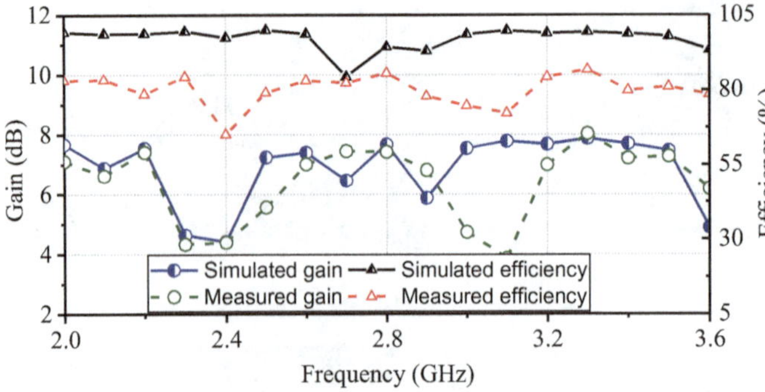

Fig. 4.18 Comparison of the simulated and measured gain and efficiency of the proposed endfire antenna. Reproduced courtesy of The Electromagnetics Academy

the relative bandwidth is about 82%. The measured results are in good agreement with the simulation results.

Figure 4.18 shows the measured gain and efficiency curves of the antenna prototype. The measured radiation efficiency of the antenna is about 80%. In addition, the measured gain of the antenna is close to the simulation. Except for 3.1 GHz, the measured gain is slightly lower than the simulation result, because the efficiency of the test is reduced at 3.1 GHz.

In addition, at 2.3 and 2.4 GHz, the gain of the antenna also decreases. This is because at these two frequencies, the electromagnetic field in the slot reverses to the SSPPs structure in the form of surface wave at the opening of the triangle, which inhibits the right transmission of electromagnetic wave in SSPPs mode. Therefore, the main radiation structure of these two frequencies is the triangular slot. So, the beam becomes wider, resulting in the decrease of gain. Figure 4.19 shows the power flow diagram of the antenna. It can be seen that the slot is the main radiator.

Therefore, it can also be said that the introduction of SSPPs structure is equivalent to modulating the EM signal in the slot waveguide structure. And it

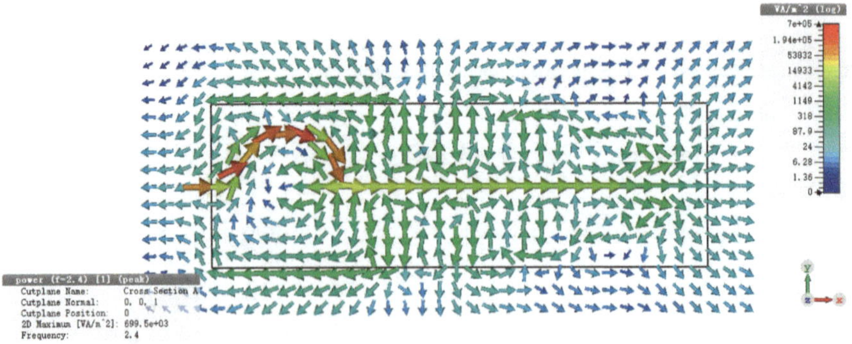

Fig. 4.19 Power flow of proposed endfire antenna at 2.4 GHz

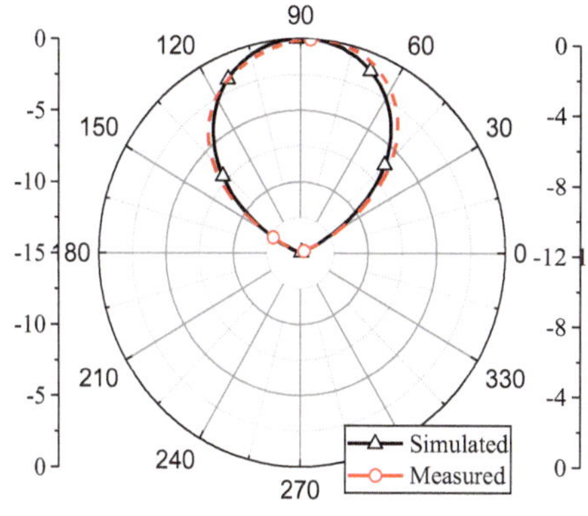

(a)

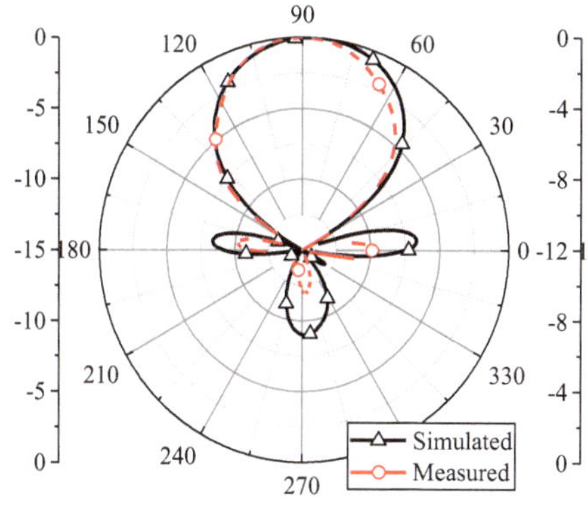

(b)

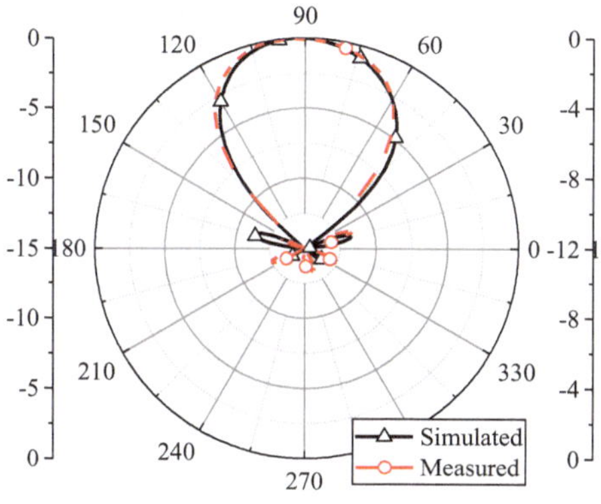

(c)

Fig. 4.20 Radiation patterns at 2.4 GHz, 2.7 GHz, and 3.4 GHz, respectively. Reproduced courtesy of The Electromagnetics Academy

Table 4.2 Performance comparison of endfire SSPPs antennas

References	Bands (GHz)	Size	Peak gain (dBi)	Radiation efficiency
[5]	9–10.5	$3.6\ \lambda_c \times 0.5\ \lambda_c$	~9.7	High
[7]	7.5–8.5	$2.9\ \lambda_c \times 0.8\ \lambda_c$	9.2	96%
[8]	6	$2.3\ \lambda_c \times 0.6\ \lambda_c$	7	92.3%
This letter	2–3.4	$1.9\ \lambda_c \times 0.6\ \lambda_c$	8	80%

Reproduced courtesy of The Electromagnetics Academy
$^*\lambda_c$ is the wavelength at the center frequency

increases the aperture of the antenna, effectively reduces the beam width, and improves the directivity and gain of the antenna.

The plane patterns of the antenna at 2 GHz, 2.7 GHz and 3.4 GHz are shown in Fig. 4.20. Radiation patterns at 2.4 GHz, 2.7 GHz, and 3.4 GHz, respectively. The simulated and measured data are in good agreement. The side lobe and back lobe of the antenna are relatively small. At 3.4 GHz, the sidelobe level in xoz plane increases.

Table 4.2 shows the performance comparison between the proposed SSPPs antenna and the existing SSPPs antenna. It can be seen that the proposed end fire antenna based on the gradual SSPPs structure has a wider impedance bandwidth, while taking into account the miniaturization and high gain. On the other hand, the simulated radiation efficiency of the antenna is 94%, and the measured radiation efficiency is 80%.

4.3 Summary of This Chapter

In this chapter, an asymmetric feed method is proposed to realize SSPPs end fire antenna, and the simulation results show that the proposed method is feasible. Furthermore, in order to improve the working bandwidth of the end fire antenna, a wideband and high efficiency SSPPs end fire antenna is proposed, and its electromagnetic field characteristics are analyzed by simulation. The experimental results show that the antenna can achieve good end fire performance in the range of 2–3.4 GHz. The measured maximum gain reaches 8 dBi, and the average efficiency is 80% over the operating band. Ease of fabrication, low profile, and high efficiency render this antenna highly suitable for endfire applications.

References

1. A. Kianinejad et al., Design and modeling of spoof surface plasmon modes-based microwave slow-wave trans-mission line. IEEE Trans. Microw. Theory Tech. **63**(6), 1817–1825 (2015)
2. S.K. Gu, H.F. Ma, B.G. Cai, T.J. Cui, Continuous leaky-wave scanning using periodically modulated spoof plasmonic waveguide. Sci. Rep. **6**, 29600 (2016)

3. Q. Zhang, Q. Zhang, Y. Chen, Spoof surface plasmon polariton leaky-wave antennas using periodically loaded patches above PEC and AMC ground planes. IEEE Ant. Wirel. Propag. Lett. **16**, 3014–3017 (2017)
4. D. Liao, Y. Zhang, H.G. Wang, Wide-angle frequency-controlled beam scanning antenna fed by standing wave based on the cut-off characteristics of spoof surface plasmon polaritons. IEEE Ant. Wirel. Propag. Lett.
5. Y.J. Han et al., 360° scanning multi-beam antenna based on spoof surface plasmon polaritons. Acta. Phys. Sin. **65**(14) (2016)
6. K. Zhuang, J.P. Geng, Z. Ding et al., A compact endfire radiation antenna based on spoof surface plasmon polaritons in wide bandwidth. Prog. Electromag. Res. M. **79**, 147–157 (2019)
7. A. Kandwal, Q. Zhang, X. Tang, L.W. Liu, G. Zhang, Low-profile spoof surface plasmon polaritons traveling-wave antenna for endfire radiation. IEEE Ant. Wirel. Propag. Lett. **17**(2), 184–187 (2017)
8. J.Y. Yin, D. Bao, J. Ren, H.C. Zhang, B.C. Pan, Y. Fan, T.J. Cui, Endfire radiations of spoof surface plasmon polaritons. IEEE Ant. Wirel. Propag. Lett. **16**, 597–600 (2017)

Chapter 5
Low Frequency Omnidirectional Antenna Based on Spoof Surface Plasmon Polaritons

Junping Geng, Dong Wang, and Jiawei Han

5.1 Introduction

Low frequency communication system is still the main means of many special communication, in which RF and antenna account for 80% of the total system volume. On the one hand, people's demand for large capacity and high rate of communication is increasing, which puts forward higher requirements for the performance of the antenna. When the size of the traditional antenna is close to the working wavelength, it is possible to obtain better radiation characteristics. On the other hand, electronic equipment miniaturization is the trend of wireless communication system development, especially the antenna as an important part of the system, is the key to the system miniaturization. How to realize miniaturization and high-performance low-frequency antenna has become the bottleneck of low-frequency communication system. In this chapter, SSPPs structure is introduced into the low-frequency band to design the antenna, and the ideas and methods to solve the problem of miniaturization, low profile and high-performance low-frequency antenna are given.

5.2 A High-Efficiency Broadband Omnidirectional UHF Patch Antenna Applying Surface Plasmon Polaritons for Handheld Terminals

In this section, SSPPs corrugated metallic structures [1] is used as the main radiation element to realize high efficiency transmission, a split SSPPs ring, and a spiral SSPPs scatter are introduced to be load to improve the surface currents distribution

J. Geng (✉) · D. Wang · J. Han
Department of Electronic Engineering, Shanghai Jiao Tong University, Shanghai, China
e-mail: gengjunp@sjtu.edu.cn

J. Geng et al., *Spoof Surface Plasmon Polaritons Antenna*, Modern Antenna,
https://doi.org/10.1007/978-981-16-4721-5_5

to better omnidirectional radiation performance of the antenna [2]. The proposed SSPPs antenna can work in UHF band with subwavelength resonance state, and it radiates omnidirectional pattern without additional large ground. In fact, higher efficiency of SSPPs mode will further increase the radiation efficiency of this SSPPs antenna. In another side, the characteristics of the SSPPs structure in the antenna can improve the performances of wide band, high gain, and miniaturization simultaneously, so that the SSPPs antenna shows high radiation efficiency, broad band, miniaturization, and omnidirectional characteristics, which is convenient to be used in modern wireless terminal systems. Process of the antenna design and the performance will be discussed in detail in the following.

Here CST Microwave Studio is used to simulate the electromagnetic performances of the antenna.

5.2.1 *Antenna Geometry Structure and Analysis*

A. *Antenna Geometry Structure*

The structure of the antenna is shown in Fig. 5.1. The details of the transition part (the second part) are shown in Fig. 5.1b, and the decomposition of each part of the antenna is shown in Fig. 5.1c.

The metal structure of the antenna is on the same plane and printed on the FR4 dielectric plate. The thickness of the FR4 dielectric plate is 1 mm and the dielectric

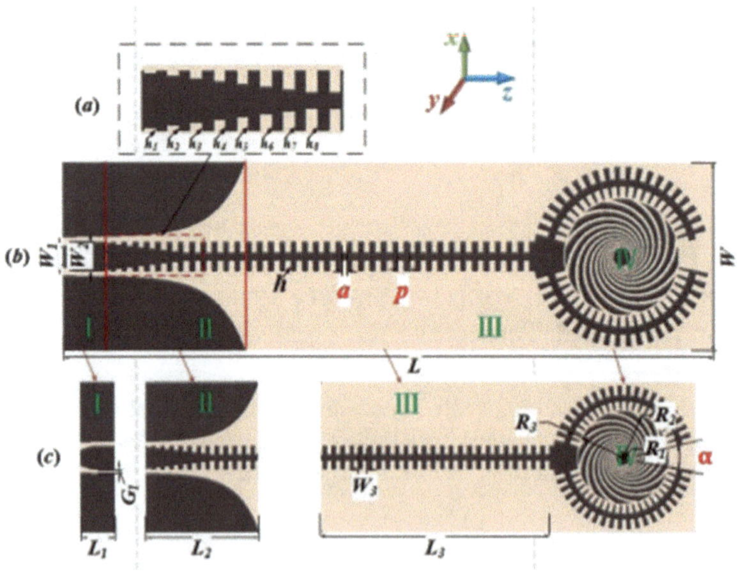

Fig. 5.1 The geometry of the SSPP patch antenna. **a** Bridge part; **b** antenna model; **c** each part model. Figure reproduced with permission from Ref. [2], © 2018 IEEE

constant is 4.3. The antenna consists of four parts. Part I is CPW structure, which is composed of a microstrip line with initial width of W_1 gradually changing to width of W_2 and ground on both sides. Part II is transition structure, which is a microstrip line with ripple structure, whose ripple depth gradually transits from h_1 to h_2, where the gradient is 0.5 mm; Part III is the transmission and radiation part, which is a corrugated microstrip line with the length of L_3 and the depth of $h_8 = 4$ mm; Part IV is the loading structure, which is composed of a ring corrugated metal microstrip line with a notch (cutting angle α, here we chose $\alpha = 20°$) and a plane spiral shape SSPPs scattering structure, in which the SSPPs scattering structure consists of twelve equiangular spiral microstrip lines surrounding a circular patch. A small slot ($S = R_3 - R_2$, here we chose $S = 0.3$ mm) is cut between the SSPPs ring and SSPPs scatter. Part IV mainly compensates the phase for standing wave and adjusts the fields distribution to be omnidirectional. SMA connects the CPW for side feed, and the overall size of the antenna is 230 mm × 65 mm × 1 mm.

B. *Principle of High Efficiency and Miniaturization*

As shown in Fig. 5.1, the UHF antenna consists of four parts. The functions of each part are different. Firstly, Part I is the traditional CPW structure, which mainly receives the electromagnetic energy from SMA connector efficiently, and transmits it to the transition structure. The transition bridge structure transforms the transmitted electromagnetic signal from waveguide mode to SSPPs mode. In the first part, the electric field propagates in the traditional waveguide mode, and the electric field is tightly bound in the gap between the microstrip line and the ground. When the electric field enters the second part, with the gradual change of the fan-shaped ground, the binding of the fan-shaped ground to the electric field gradually weakens, and the electrical field line gradually bends. In this part, the waveguide mode coexists with the SSPPs mode. When the electric field enters the third part, the SSPPs mode completely disappears in this case, the electric field is periodically distributed as a semi closed field line. Finally, a small amount of remaining electromagnetic wave enters the fourth part. As shown in Fig. 5.2a, the power flows in the center part at 320 MHz along the transmission part, and forks along the split SSPPs ring, and the power flows away from the center SSPPs part symmetrically—up and down, left and right, and almost is omnidirectional in the horizontal plane (*xoy* plane). Certainly, the fourth part is mainly to compensate the phase of standing wave and prevent electromagnetic energy overflowing from the top.

It is noted that the transmission efficiency of SSPPs wave guide is larger than microstrip. In Fig. 5.2a, terminal load part makes the power flow distribution vertical part uniform, so that the power flow away from the antenna is mainly in horizontal omnidirectional radiation. In the transition part, when the ripple depth of SSPPs structure changes from 0 to 4 mm, the corresponding propagation constant in x direction will also vary from k_0 to k_x, realizing the perfect matching from CPW structure (k_0) to SSPPs structure (k_x). The binding effect of the structure on the electromagnetic beam can be expressed by formula (5.1).

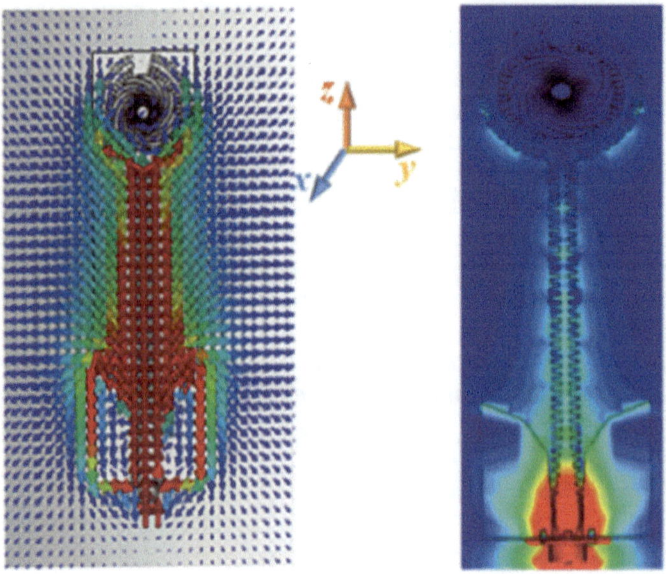

(a) (b)

Fig. 5.2 Power flow and electric field distribution of the proposed antenna at x = 0 plane. **a** Power flow distribution at 320 MHz; **b** electric field distribution at 320 MHz. Figure reproduced with permission from Ref. [2], © 2018 IEEE

$$\frac{\lambda_p}{\lambda_0} = \frac{1}{\sqrt{1 + \frac{a^2}{p^2}\tan(k_0 h)}} \tag{5.1}$$

Here, λ_0 is the wavelength of the electromagnetic wave in the vacuum. λ_p is the wavelength of the electromagnetic wave in the SSPPs mode. From Eq. (5.1), the ratio of λ_p/λ_0 is about $0.915 < \lambda_p/\lambda_0 < 0.925$ during 300–400 MHz, and even stronger binding to electromagnetics wave in the higher frequency band. Figure 5.2b is the electric field distribution at 320 MHz, which means that the proposed UHF SSPPs antenna is an efficient structure to support and propagate SSPPs wave.

5.2.2 *Parameter Analysis*

From the previous simulation, we can see that the ripple depth (h) of the corrugated metal microstrip line can effectively affect the reflection coefficient of the antenna.

The simulation results of $|S_{11}|$ parameters and radiation efficiencies of the antenna with three different ripple depths h (4, 2, 0 mm) are shown in Fig. 5.3. It can be seen that with the increase of h, the impedance bandwidth is 0.305–0.332 GHz, 0.328–0.360 GHz and 0.341–0.372 GHz respectively, which indicates that the working frequency band of the antenna gradually moves to the low frequency region, and the $|S_{11}|$ performance of the antenna is obviously improved. Figure 5.3 also shows the radiation efficiency of the antenna. It can be seen that when h = 2, 4 mm, the radiation efficiencies of the antenna are greater than 90% in the available impedance bandwidth, and the highest values are 95.5% and 98.7% respectively. When h = 0, the middle part of the structure becomes a pure microstrip line, and the radiation efficiency decreases, with the peak only 91.2%.

In addition, because the end loading has a great influence on the phase of standing wave and the electromagnetic wave overflowing from the top, the influence of the radius R_2 of the spiral loading structure and the notch angle (α) of SSPPs ring on the antenna performance are discussed here.

Figure 5.4 shows the $|S_{11}|$ results affect by the spiral loading structure and SSPP ring slot s ($s = R_1 - R_2$, s increase means R_2 decrease). It can be seen that with the decrease of R_2, the working frequency band of the antenna will shift to higher frequency, but when R_2 decreases to a certain value, the $|S_{11}|$ of the antenna nearby does not change, because when R_2 is very small, the constraint effect of the gap s on electromagnetic field almost disappears when the gap width between spiral arm and SSPP ring increases to a certain extent.

Figure 5.5 shows the reflection coefficient $|S_{11}|$ of the antenna varying with the notch angle of SSPP corrugated metal ring. It can be seen that with the increase of the notch angle, the electrical length of the antenna will decrease, resulting in the working frequency shift to high frequency.

The optimal parameters of the optimized antenna are shown in Table 5.1. The overall length of the antenna is 0.169 λ_0, and the radiation efficiency of the antenna at the central frequency point is 98.7%, which indicates that the antenna achieves miniaturization and high efficiency.

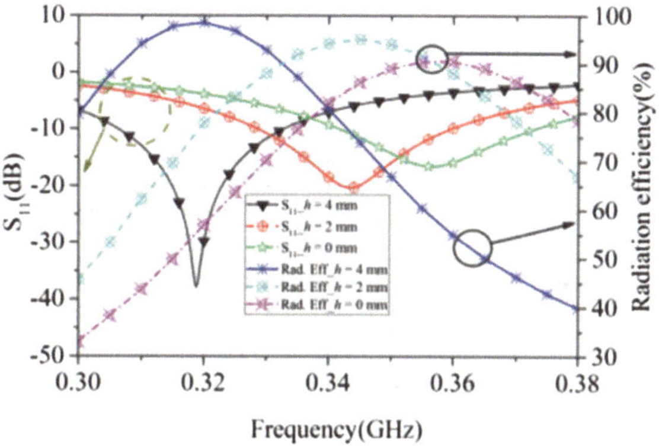

Fig. 5.3 Comparison of the simulated $|S_{11}|$ values and radiation efficiency for h = 0, 2, 4 mm. Figure reproduced with permission from Ref. [2], © 2018 IEEE

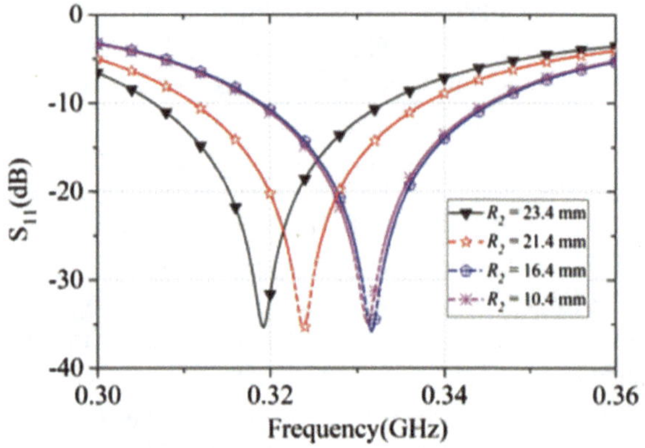

Fig. 5.4 Comparison of the simulated $|S_{11}|$ values for the R_2 = 10.4, 16.4, 21.4 and 23.4 mm. Figure reproduced with permission from Ref. [2], © 2018 IEEE

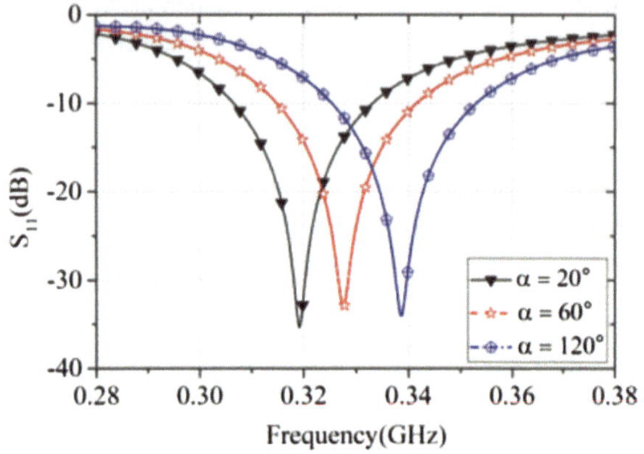

Fig. 5.5 Comparison of the simulated $|S_{11}|$ values for the R_2 = 10.4, 16.4, 21.4 and 23.4 mm. Figure reproduced with permission from Ref. [2], © 2018 IEEE

Table 5.1 Optimized parameters of the proposed antenna

Parameter	L	L_1	L_2	L_3	W	W_1	W_2
Size (mm)	230	15.2	45.8	104.2	65	6	10
Parameter	W_3	R_1	R_2	R_3	S	h	
Size (mm)	4	3.5	20.3	20.6	0.3	4	

Table reproduced with permission from Ref. [2], © 2018 IEEE

5.2.3 Experiments Results and Discussion

A. *Reflection Coefficient* $|S_{11}|$ *and Realized Gain*

According to the best parameters in Table 5.1, the prototype of antenna is processed and tested. The real antenna is shown in Fig. 5.6.

Figure 5.7 shows the reflection coefficient $|S_{11}|$ curves of the antenna measured by Agilent N8163a vector network analyzer. The measured impedance band of $|S_{11}| < -10$ dB is 302–335 MHz. The measured results are basically consistent with the simulation results.

Fig. 5.6 The photo of the antenna material object. Figure reproduced with permission from Ref. [2], © 2018 IEEE

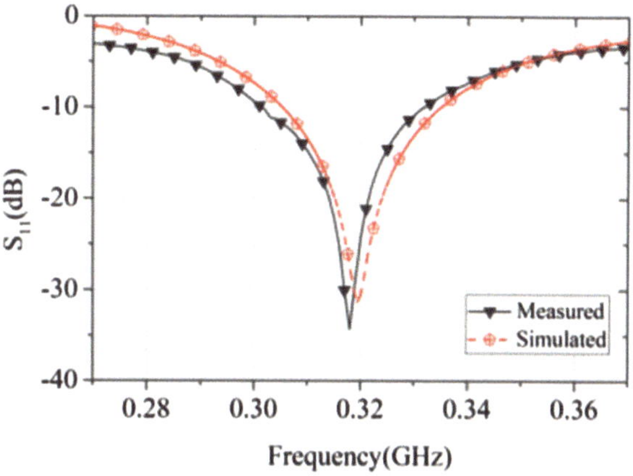

Fig. 5.7 The simulated and measured $|S_{11}|$ of the antenna. Figure reproduced with permission from Ref. [2], © 2018 IEEE

The actual measured gain is shown in Fig. 5.8 and the trend of the measured result is basically consistent with the simulation result. The maximum measured gain is 1.76 dBi, which appears at 320 MHz, and the maximum error is 0.98 dbi at 328 MHz. When the frequency is lower than 330 MHz, the measured gain is less than the simulation value. This may be due to machining error, dielectric constant

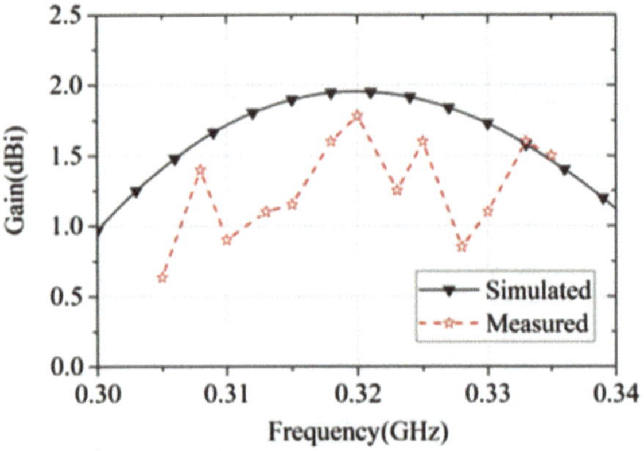

Fig. 5.8 Simulated and measured realized gains of the proposed antenna. Figure reproduced with permission from Ref. [2], © 2018 IEEE

error and other physical reasons. On the other hand, because of the large size of the antenna transmitting source at UHF band, a bracket is needed to support the transmitting signal source in the microwave anechoic chamber in the practical measurement, which will bring the corresponding electromagnetic wave reflection and may cause errors in the test. Although the test results are inconsistent with the simulation results in some directions, the main radiation direction is basically consistent with the simulation results.

B. *Radiation Performance*

The far-field radiation characteristics of the antenna are measured in a microwave anechoic chamber. Figure 5.9 shows the E-plane (xoz) and H-plane (YOZ) patterns tested at 310, 320 and 330 MHz. The main lobe beam direction of the pattern is basically consistent with the simulation results.

Figure 5.9 also shows the measured cross polarization pattern. The cross polarization is 16 dB smaller than the main polarization in the range of main radiation direction. The measured results show that the antenna has good omnidirectional radiation characteristics in the horizontal plane. The measured 3 dBi beamwidth of the antenna on the E-plane is about 84°–112.5° which is much higher than that of the traditional half wave oscillator.

Table 5.2 shows the performance comparison between our designed antenna and the existing UHF antenna [3–6]. Compared with the Refs. [3–6], the gain of our multi band antenna is slightly lower, but the relative size of the antenna is much smaller than that of the reference antenna. It can be seen that compared with the traditional UHF antenna, the miniaturized low profile broadband high efficiency UHF antenna based on SSPPs can give good consideration to the miniaturization, high efficiency and high gain performance of the antenna, which is more suitable for application in modern wireless communication system.

In this section, a high-efficiency broadband omnidirectional UHF patch antenna applying SSPPs for handheld terminals is studied. This high-efficiency antenna balanced high gain, wide band, omnidirectionality, and miniaturization. The SSPPs' strip is not only transition but also radiation. The work band is sensitive to the depth

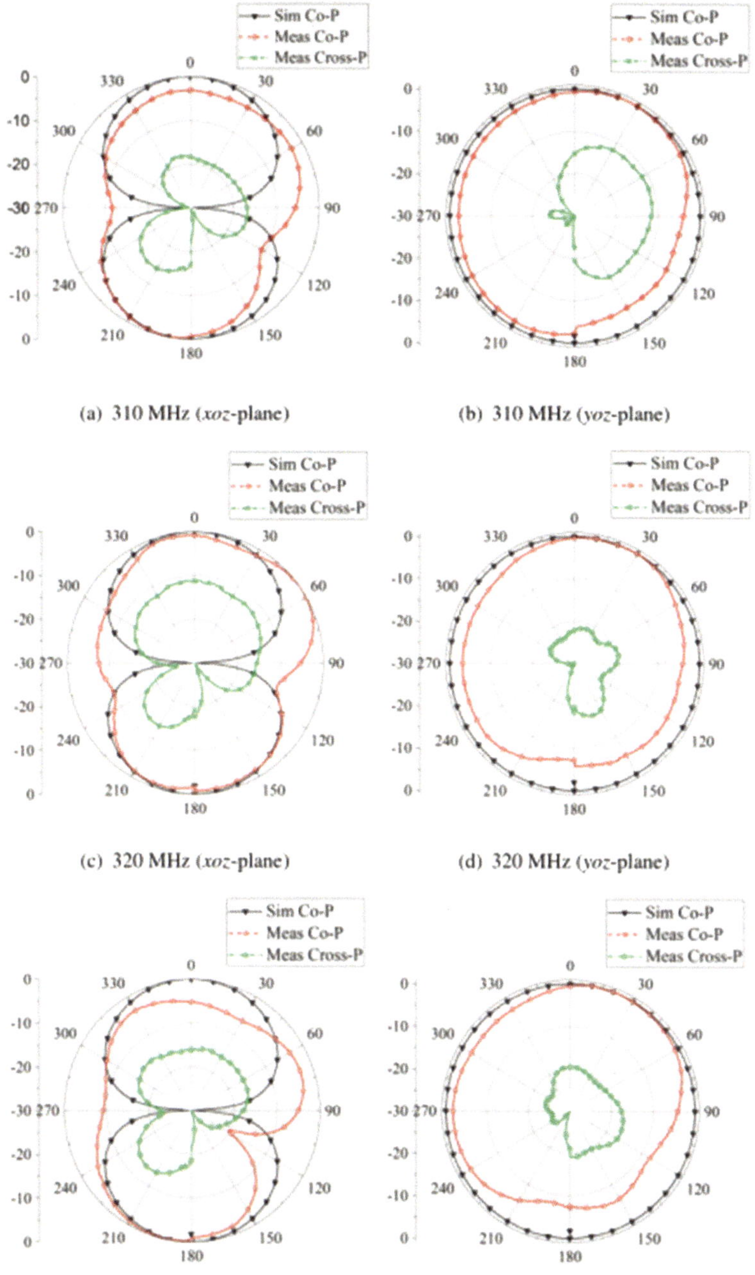

(a) 310 MHz (*xoz*-plane)

(b) 310 MHz (*yoz*-plane)

(c) 320 MHz (*xoz*-plane)

(d) 320 MHz (*yoz*-plane)

(e) 330 MHz (*xoz*-plane)

(f) 330 MHz (*yoz*-plane)

Fig. 5.9 Simulated and measured radiation patterns of the proposed antenna (E-plane: $\varphi = 0°$, H-plane: $\varphi = 90°$). Figure reproduced with permission from Ref. [2], © 2018 IEEE

Table 5.2 Performance comparison of UHF band antennas

References	Bands (GHz)	Size	Peak gain (dBi)	Radiation efficiency	Complexity
[3]	0.354–0.360	$0.7\lambda \times 0.7\lambda \times 0.025\lambda$	3	High	High
[4]	0.388–0.395	$1.4\lambda \times 0.3\lambda \times 0.105\lambda$	6.7	High	Moderate
[5]	0.3–0.87	$0.3\lambda \times 0.3\lambda \times 0.08\lambda$	4.15	High	High
[6]	Around 0.4 GHz	$0.1\lambda \times 0.1\lambda \times 0.014\lambda$	−6	Low	High
This letter	0.302–0.335	$0.26\lambda \times 0.065\lambda \times 0.001\lambda$	1.76	98.7%	Low

of the comb strip, cut angle of the split SPPs' ring, and the slot between the split SPPs' ring and the spiral arms. The measured impedance bandwidth is about 10%, covering 0.302–0.335 GHz. The realized gain measured over this operating bandwidth ranged from 0.9 to 1.76 dBi. It is noted that by using SPPs transition line and a loading SPPs split ring with conventional SPPs scatter, high radiation efficiency (more than 90%) and omnidirectional radiation can be realized. The radiation patterns of the proposed antenna show good, stable, and omnidirectional characteristics in the horizontal plane and wider band beam (84°–112.5°) in the E-plane. Consequently, it easily meets most mobile wireless communication terminal system requirements.

5.3 UHF Antenna with Omnidirectional Vertical Polarization

Omnidirectional vertical polarization antenna is widely used. With the development of miniaturization and the increasing shortage of spectrum resources, omnidirectional miniaturization multi band antenna has become a research hotspot at home and abroad [7, 8].

In Refs. [9–15], a number of omnidirectional low profile antennas are designed by using different technologies and methods.

The geometry of the proposed low-profile dual-band vertically polarized omnidirectional antenna is shown in Fig. 5.10 [16, 17]. The antenna mainly comprises an upper circular patch with annular slots, a metallic cone, twelve cylindrical shorting pins and a circular ground plane. It is fabricated on a 60 mm thick circular shaped FR4 substrate with relative permittivity of 4.4 and loss tangent of 0.02. The 50-Ω SMA is connected with ground plane to feed the proposed antenna. By connecting cylindrical shorting pins from the upper patch to the ground plane, size of the structure is largely reduced. The diameter of the circular patch is about 0.413 the free-space wavelengths and the antenna height is about 0.0138 the free-space wavelengths of the low operating frequency. Furthermore, to achieve the dual-band characteristics, the two annular slots on the upper patch is introduced.

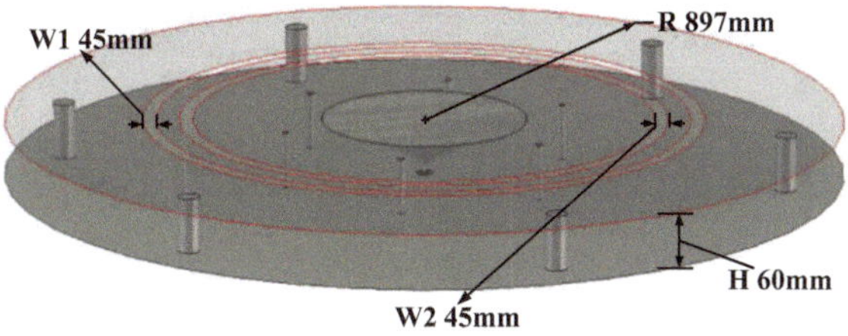

Fig. 5.10 Geometry of the proposed antenna. Figure reproduced with permission from Ref. [16], © 2020 IEEE

The proposed antenna is fabricated and simulated in CST Microwave Studio. The simulated result of S parameters is shown in Fig. 5.11. The reflection coefficient is below −10 dB with two frequency bands from 69 to 72 MHz and 317 to 347 MHz, and the relative bandwidths of the lower and upper bands can reach 4.3 and 9%.

The radiation patterns for the proposed antenna are simulated. Figures 5.12a and 5.13a show the simulated E-plane radiation patterns at 72 MHz and 320 MHz, respectively. And the H-plane radiation patterns are shown in Figs. 5.12b and 5.13b. According to the results, the antenna clearly shows an omnidirectional radiation pattern at all frequencies.

In this paper, a low-profile dual-band vertically polarized omnidirectional antenna is proposed. Low profile of the antenna height about 0.0138 free-space wavelengths of the low operating frequency is achieved by cylindrical shorting pins, compared with 0.07 in [10] and 0.05 in [13]. And two annular slots are also introduced to realize the dual-band characteristics. In the two operating frequency bands, the radiation patterns exhibited omnidirectional radiation property which makes the antenna suitable for future vehicular mobile wireless applications.

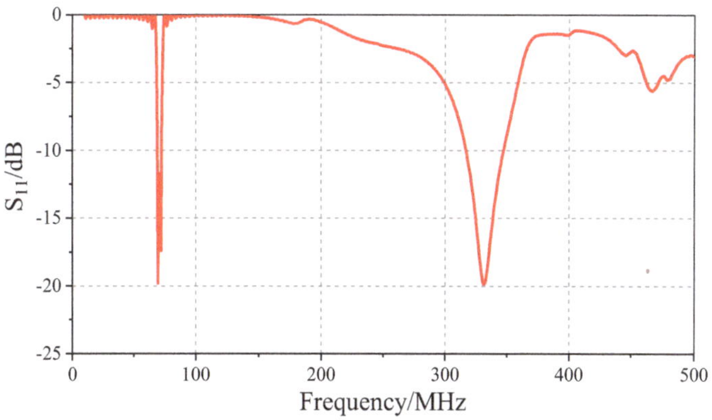

Fig. 5.11 Simulated S parameters of the proposed antenna. Figure reproduced with permission from Ref. [16], © 2020 IEEE

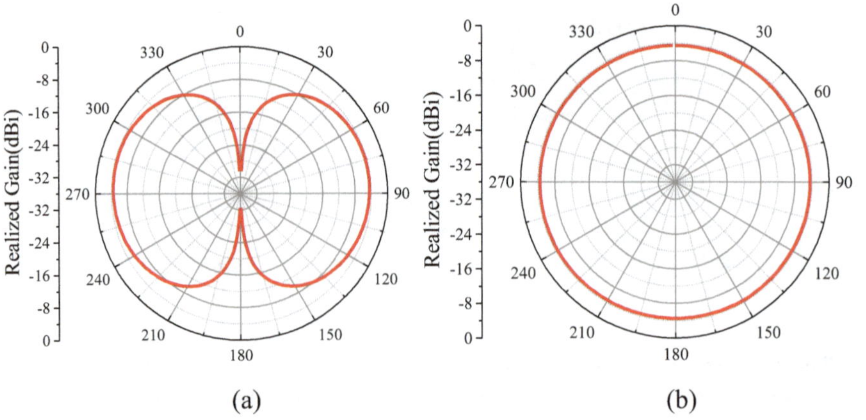

Fig. 5.12 Simulated radiation pattern at 72 MHz. a E-plane; b H-plane. Figure reproduced with permission from Ref. [16], © 2020 IEEE

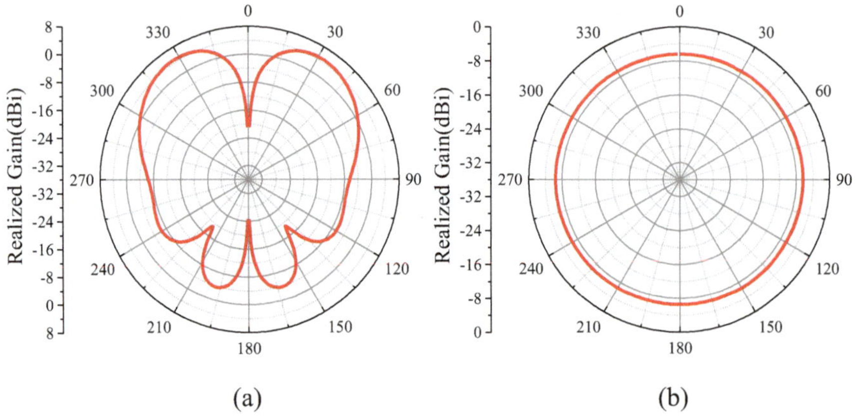

Fig. 5.13 Simulated radiation pattern at 320 MHz. a E-plane; b H-plane. Figure reproduced with permission from Ref. [16], © 2020 IEEE

5.4 UHF Omnidirectional Antenna with Vertical Polarization

With the quickly developing of mobile wireless communication, the omnidirectional antenna is very needed for various applications, such as RFID, broadcasting, vehicular communication and so on. Whip antenna has a good omnidirectional radiation pattern, but with low gain, especially too high profile in low frequency band. Recently, the conformal antenna is proposed and developed rapidly., and some conformal antenna has been used on the shell of the vehicle/airborne, but the strength of these conformal antennas is poor in bearing load.

In the past years, spoof surface plasmon polaritons (SSPPs) structure has been the research focus for its high transmission efficiency, subwavelength and available structure, which supports different mode electromagnetic waves propagating along the corrugated metal surface similar to SPPs in optics [14–20]. Different from the conventional transmission line, the SSPPs structure shows deep field high confinement and adjustable dispersion characters, which has good potential to microwave devices and antenna. A high-efficiency broadband omnidirectional UHF patch antenna based on SSPPs for handheld terminals is presented in [2], which provide broadside radiation by the even mode of SSPPs. In [18], a pattern reconfigurable antenna based on spoof surface plasmon polaritons transmission line with switches of PINs is proposed. It supports the switch between a broadside radiation beam in the even mode and an end-fire radiation beam in the odd mode by adjusting the status of three PIN diodes.

In this section, a vertical polarization SSPPs antenna is designed by asymmetric fed, and the omnidirectional antenna with vertical polarization is constructed by ring array of the SSPPs antenna [21]. Later, the improved SSPPs antenna with half profile is presented and combine the ring array too.

5.4.1 Study on Low-Profile Spoof Surface Plasmon Polaritons End-Fire Antenna with Vertical Polarization and Its Annular Array

The main idea of this design is to realize the horizontal omnidirectional antenna by SSPPs end-fire antenna. The original end-fire antenna has the characteristics of low profile and vertical polarization, and then uses the end-fire characteristics to realize omnidirectional radiation through the combination of multiple antennas.

5.4.1.1 Low Profile SSPPs End-Fire Antenna with Vertical Polarization

For the double arm SSPPs structure, when the odd mode traveling wave is excited on it, the end-fire radiation can be realized. If the arms of SSPPs are placed vertically, vertical polarization can be achieved. This section designs an initial model of a vertical polarized SSPPs end fire antenna fed by monopole, which is denoted as Antenna #1, as shown in Fig. 5.14. The antenna consists of two parts, monopole feed structure and double arm SSPPs transmission and radiation structure. The SSPPs structure is composed of three parts. The left part I is a gradual structure, which belongs to the transition part. It is mainly convenient for the monopole to efficiently couple the electromagnetic energy into the SSPPs structure. The middle part II is the SSPPs transmission waveguide, which mainly binds the electromagnetic wave in the SSPPs structure, so that it can be transmitted to the right by

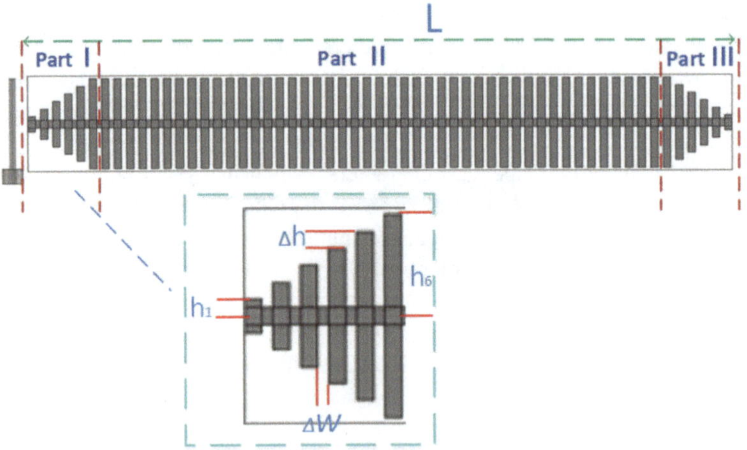

Fig. 5.14 Geometry of the antenna #1

traveling wave. The right part III is also a gradual structure, which is mainly used to realize the impedance matching between traveling wave and free space on the end of SSPPs structure, so as to facilitate the efficient radiation of electromagnetic wave to free space and realize end fire.

The return loss curve of the antenna feed port is shown in Fig. 5.15. The impedance matching curve of Antenna #1 is given, from which it can be seen that the frequency band of $|S_{11}| < -10$ dB is 1.48–1.8 GHz, and the relative bandwidth is 19.5%. When $f = 1.6$ GHz, the electric field distribution of the antenna in the *xoy* plane is shown in Fig. 5.16a. It can be seen that the electric field distribution is anti symmetric, which shows a typical odd mode feature. The far-field pattern of the antenna at f = 1.6 GHz is shown in Fig. 5.16b, which is obvious end fire radiation.

SSPP structure itself is a broadband structure, but the monopole feed structure is a narrow-band resonant structure, resulting in the relative bandwidth of antenna #1 being 19.5%. In order to further broaden the impedance bandwidth of the antenna, the monopole in antenna #1 is replaced by a cone-like monopole, as shown in Fig. 5.17, and the whole antenna is called antenna #2. Taking the bus of the cone-like monopole as sin function can increase the effective current length of the monopole, and the current distribution is more uniform, which is also conducive to broaden the impedance matching bandwidth of the antenna. The impedance curve of antenna #2 is shown in Fig. 5.18. The frequency band of $|S_{11}| < -10$ dB is 1.22–2.08 GHz, and the relative bandwidth is 52%.

In addition, combined with the actual application environment, the antenna should have a certain carrying capacity, so the end fire antenna should be further optimized. Replace the planar SSPPs structure in antenna #2 with the columnar SSPPs structure, as shown in Fig. 5.19, and record it as antenna #3. Figure 5.20 shows the return loss curve of antenna #3. It can be seen that the frequency band of $|S_{11}| < -10$ dB of antenna #3 is 1.18–2.1 GHz, and the relative bandwidth is 56.1%.

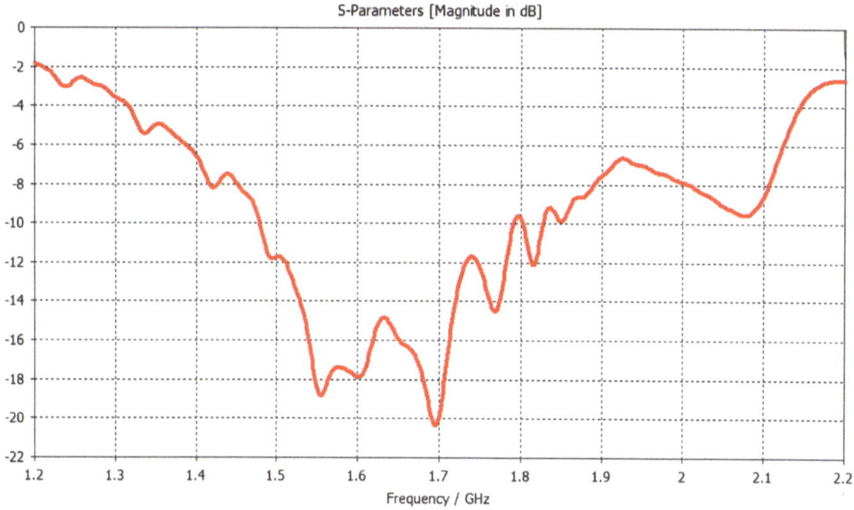

Fig. 5.15 The S-parameter of antenna #1

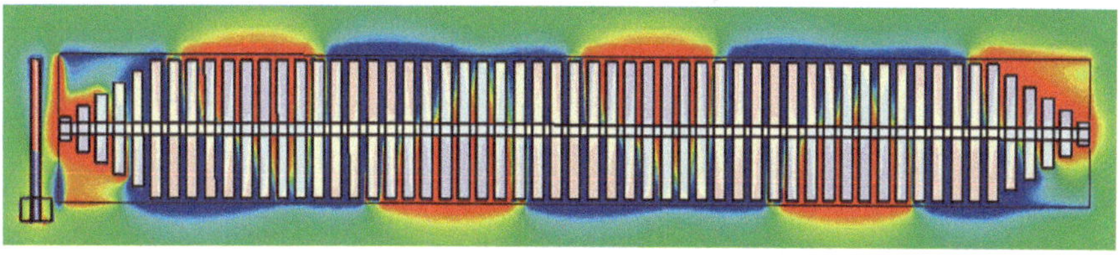

(a)

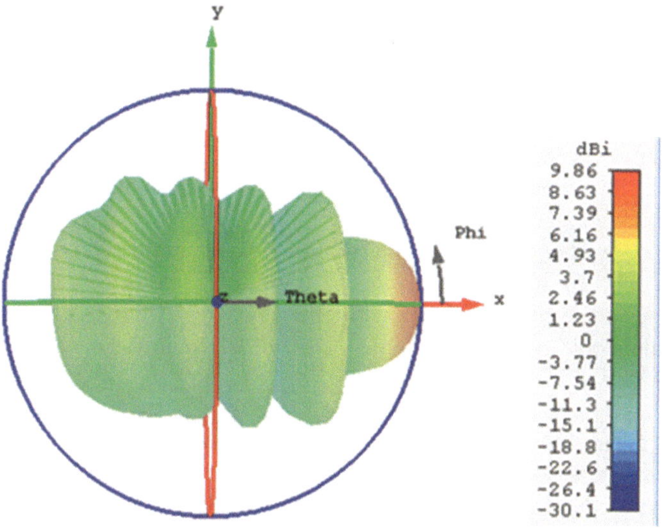

(b)

Fig. 5.16 The electric field distribution and far-field pattern of antenna #1

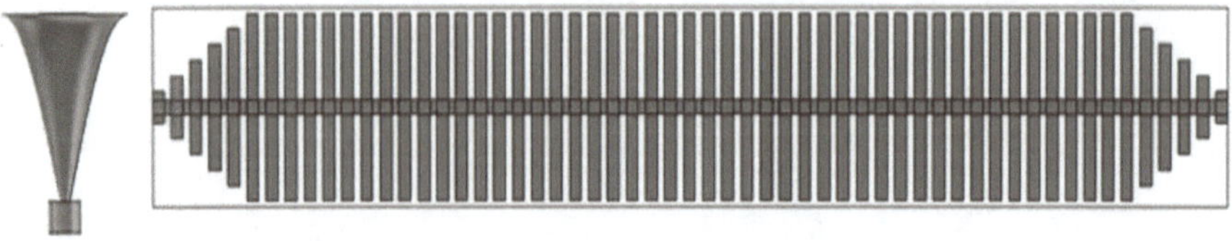

Fig. 5.17 Geometry of the antenna #2

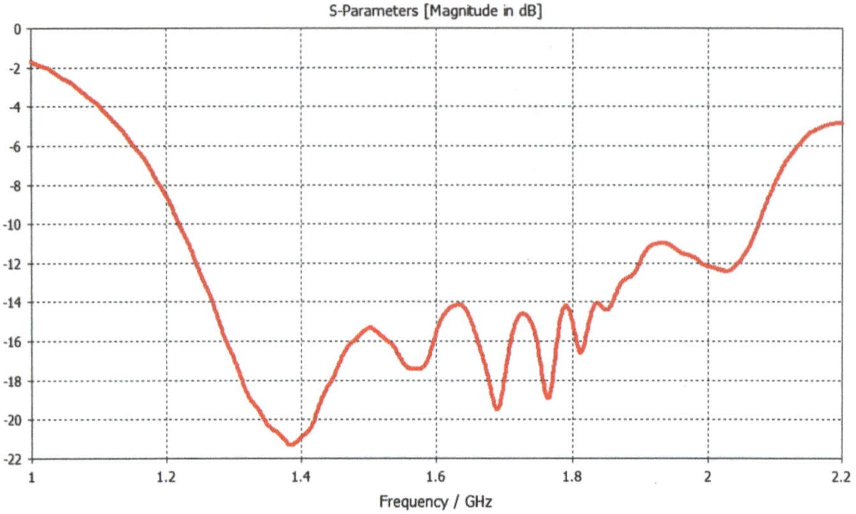

Fig. 5.18 The S-parameter of antenna #2

5.4.1.2 Low Profile Omnidirectional Antenna with Vertical Polarization Based on SSPPs Structure

In this section, a vertical polarization SSPPs antenna is designed by asymmetric fed, and the omnidirectional antenna with vertical polarization is constructed by ring array of the SSPPs antenna.

The omnidirectional antenna with vertical polarization based on SSPPs is presented and the 3D model is depicted in Fig. 5.21 Eight double-side corrugated metallic arms are evenly distributed with interval angle of 45° circle around the feed tape-like monopole. Then, the volume including all the metallic structures is filled by Fr-4 as shown in Fig. 5.21a. The complex structure mixed the metal and Fr-4 greatly strengthens the antenna's carrying capacity with small weight.

In Fig. 5.21c, the SSPPs structure includes three parts. The first part is the transition structure, the height of which varies from h_1 to h_6 with a step of Δh. The second part is the corrugated metallic arms with the same height, which is the SSPPs structure to transmit EM wave in SSPPs mode. The third part is the main

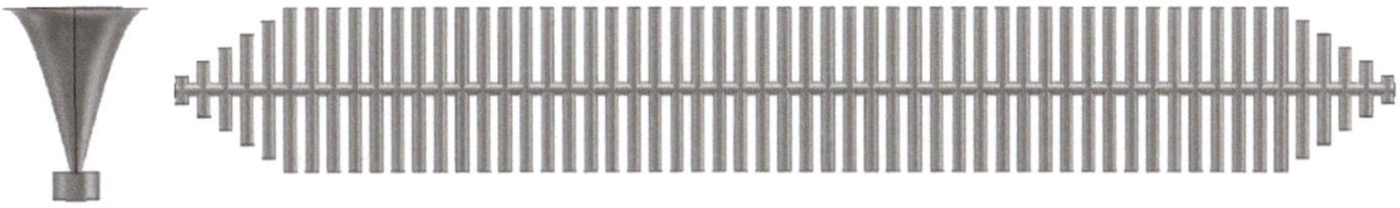

Fig. 5.19 Geometry of the antenna #3

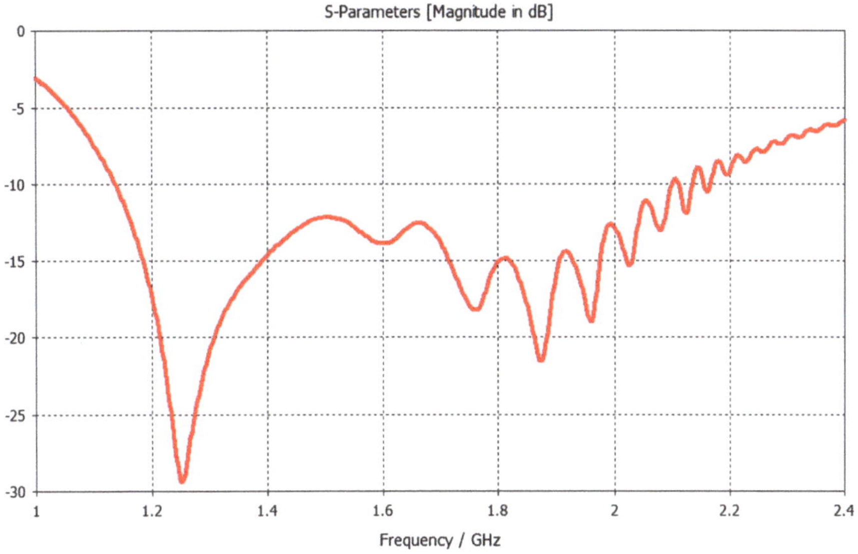

Fig. 5.20 The S-parameter of antenna #3

radiation part, and its dimension is same to that of part I. Each arm is spaced Δw apart and the whole length of the SSPPs structure is L. In Fig. 5.21b, the monopole can excite the vertical polarization wave in all direction, which can be coupled into the SSPPs structure to excite the odd mode of each double-side corrugated metallic arms, and the energy flows in radial directions along the axis poof the SSPPs structure. Therefore, eight SSPPs structures are just equivalent to eight end-fire antennas. As long as the beam width of each end-fire antenna is broad enough, the quasi omnidirectional pattern can be combined. Besides, the gain is decided by the SSPPs structure, which is proportional to the length L within certain range. When the L is large enough, the gain will keep stable.

The proposed quasi omnidirectional antenna with vertical polarization based on the SSPPs structure is designed and simulated in CST Microwave Studio. After the optimization, the final dimensions of the proposed antenna are $h_1 = 5\,\text{mm}$, $h_6 = 30\,\text{mm}$, $\Delta h = 5\,\text{mm}$, $\Delta w = 3\,\text{mm}$, and $L = 453\,\text{mm}$. Besides the height of the feed taped-like monopole is about $2 * h_6 = 60\,\text{mm}$.

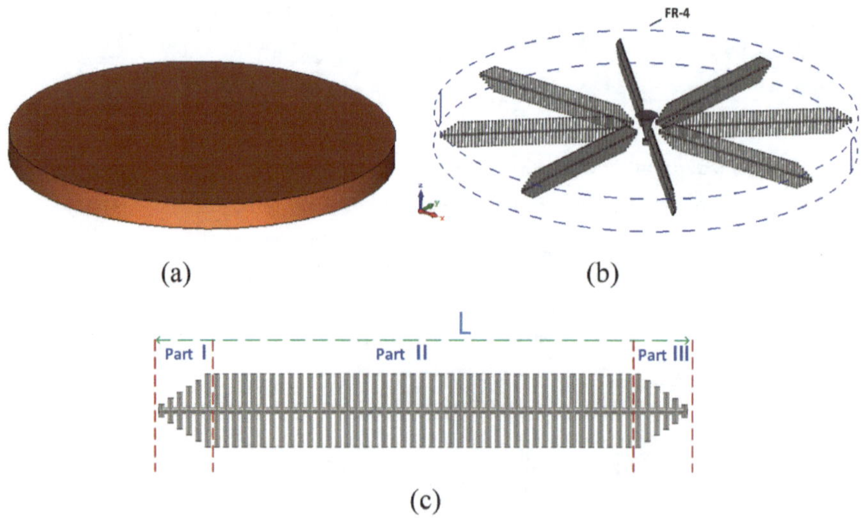

Fig. 5.21 Geometry of the proposed antenna. **a** Front view; **b** perspective view; **c** one SSPPs branch. Figure reproduced with permission from Ref. [21], © 2019 IEEE

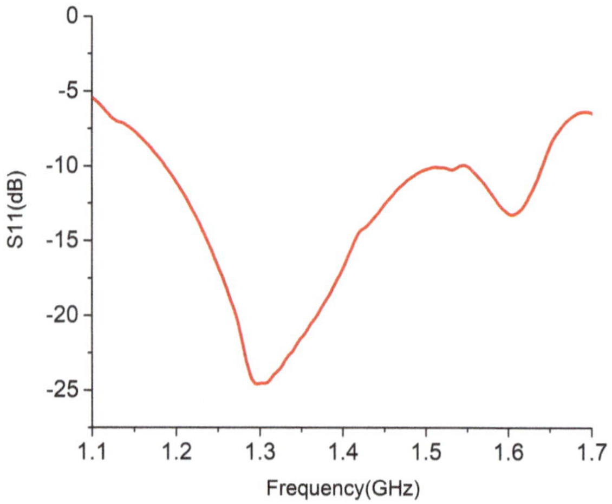

Fig. 5.22 Simulated S_{11} of the proposed antenna. Figure reproduced with permission from Ref. [21], © 2019 IEEE

The simulated return loss $|S_{11}|$ is shown in Fig. 5.22. And the impedance band with $|S_{11}| < -10$ dB is from 1.2 to 1.6 GHz, with 28.57% bandwidth. The simulated radiation pattern of the proposed antenna at 1.4 GHz is given in Fig. 5.23. It is obvious, the antenna performs a nearly omnidirectional pattern in horizontal plane.

For the composite structure, the SSPPs antenna can be realized by the combined structure including many vertical metal bars with diameters 5 mm, and the FR4 is

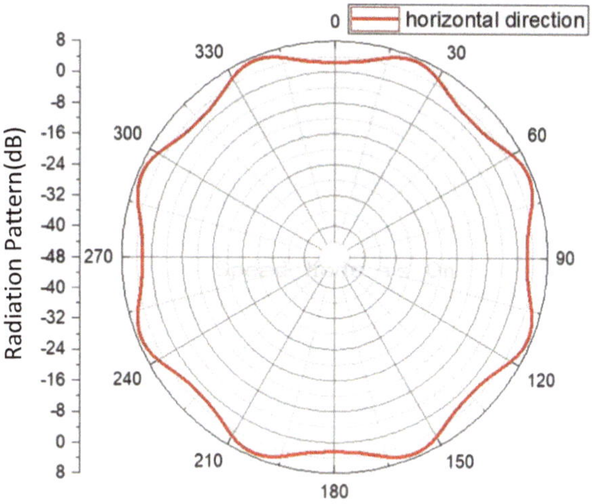

Fig. 5.23 Radiation pattern of the proposed antenna: at f = 1.4 GHz. Figure reproduced with permission from Ref. [21], © 2019 IEEE

poured in the volume. When it is solidified, it is just become a whole composite structure, so that the strength is very strong and tough but less weight. What's more, the proposed SSPPs antenna with complicated structure is easy to mount on the shell of the vehicle even takes the place of the shell with high bearing load.

This vertically polarized omnidirectional antenna based on SSPPs have a good omnidirectional radiation pattern over operating band from 1.2 to 1.6 GHz, with high gain and low profile, comparing with the monopole. And it has a good promotion in mobile vehicle communication.

5.4.2 Improved Low-Profile Spoof Surface Plasmon Polaritons Terminal Antenna and Its Annular Array

It has been shown in the previous section that the endfire antenna based the SSPPs structure can realize omni-directional radiation with vertical polarization. However, in practical applications, the effect of the floor on the antenna needs to be considered. In this section, based on the SSPPs antenna given in the previous section, we will consider the influence of the floor on the antenna, and further improve the low profile SSPPs end fire antenna and its array.

The initial end fire SSPPs antenna model (without ground) is shown in Fig. 5.24a. The original antenna adopts bilateral SSPPs structure. The SSPPs structure is printed on the dielectric plate and fed by the monopole coupling on the left side. The odd mode traveling wave is excited on the SSPPs structure. Figure 5.24b is its far field, which shows good end fire effect.

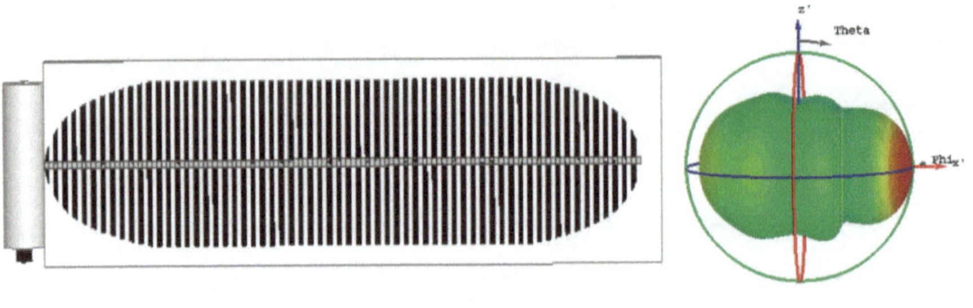

(a) (b)

Fig. 5.24 The original end-fire antenna based on SSPPs and its far-field pattern

In Fig. 5.24a, if the monopole is replaced by a symmetrical dipole, the middle symmetrical plane can be regarded as an electric wall. The floor can often be used as an electrical wall. Considering the mirror effect of the floor and the odd mode current distribution on the SSPPs structure, the original antenna can be truncated by using the complex image theory. As shown in Fig. 5.25a, the radiation characteristics of Fig. 5.25b can be simulated by using unilateral SSPPs structure as the transmission radiation structure and adding an ideal electrical boundary (infinite floor) at the bottom of the antenna. The radiation pattern is shown in Fig. 5.25b.

In practice, the ground cannot be infinite, that is to say, the ideal electrical boundary set in Fig. 5.25 cannot be realized, so the radiation characteristics of the antenna under non ideal conditions must be reconsidered. In the simulation model, the ideal boundary condition is set as a finite floor. The simulation model is shown in Fig. 5.26a, and the black line at the bottom is a finite floor. Figure 5.26b is the simulated far-field pattern of the antenna. It can be seen that the end fire pattern is warped due to the use of limited floor.

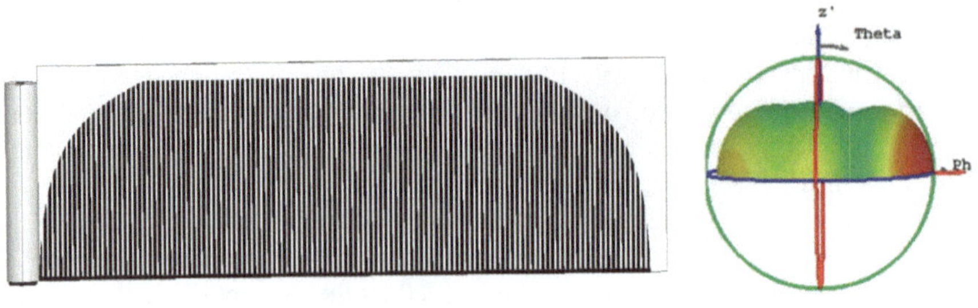

(a) (b)

Fig. 5.25 The end-fire antenna based on single-side SSPPs with PEC boundary and its far-field pattern

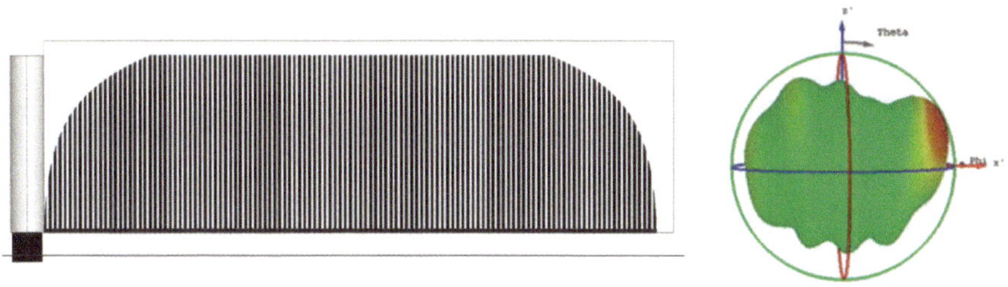

(a) (b)

Fig. 5.26 The end-fire antenna based on single-side SSPPs with ground and its far-field pattern

In order to make the main beam of antenna point to theta = 90° direction, the floor must be modified. On the basis of Fig. 5.26, bend the edge of the antenna floor downward so that the edge changes with a specific function, as shown in Fig. 5.27a. The black curve at the bottom of the figure is finite ground. Figure 5.27b shows the simulated far-field pattern of the antenna after the floor is modified. It can be seen that the main beam of the antenna with curved floor is closer to theta = 90°, and it has better end fire radiation performance. It shows that the effect of the downward bending of the limited ground edge on the upwarping of the pattern is improved, and the main beam direction is closer to theta = 90°.

Furthermore, a vertically polarized horizontal omnidirectional antenna is designed by using the end fire antenna in Fig. 5.27a. The curved floor can be expanded into a circular floor with a downward curved edge, as shown in Fig. 5.28, in which the black part is metal and the gray part is medium.

In Fig. 5.28, eight SSPPs end fire antennas are radially arranged around the center, the center of which is fed by monopole antenna coupling, and the bottom of the monopole antenna is connected with SMA connector for RF excitation feeding.

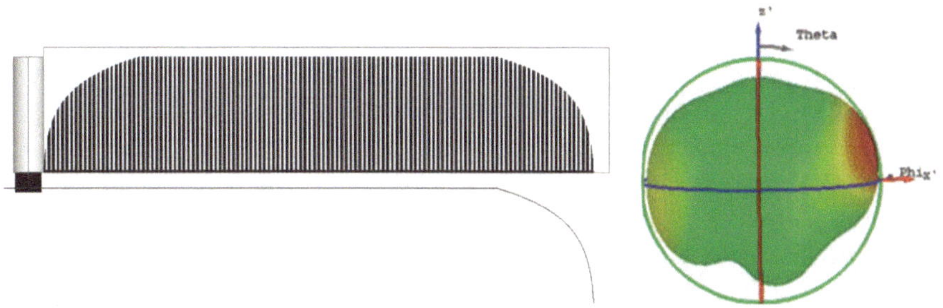

(a) (b)

Fig. 5.27 The end-fire antenna based on single-side SSPPs with improved ground and its far-field pattern

Fig. 5.28 The omni-directional antenna based on single-side SSPPs with improved ground

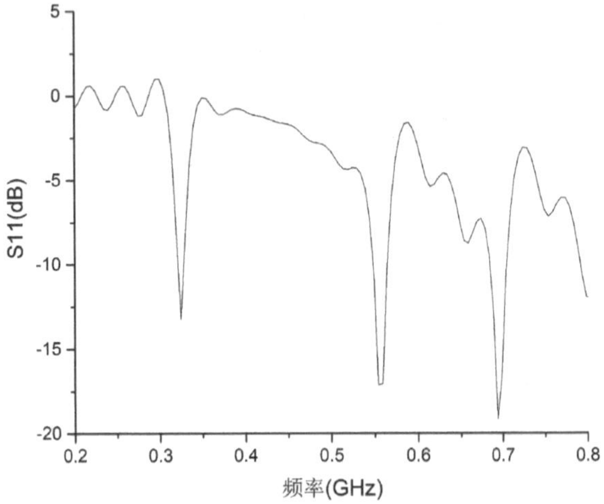

Fig. 5.29 S-parameters of the omni-directional antenna based on single-side SSPPs with improved ground

The comb like SSPPs microstrip line is etched on the FR4 plate with a dielectric constant of 4.3. The thickness of the dielectric is 1.2 mm, and the overall size of the antenna is Φ 616 mm × 60 mm. The electromagnetic signal enters from the SMA to excite the monopole antenna, then excites the eight SSPPs antennas by coupling, and finally radiates out through the end of the unilateral SSPPs structure. Because of the combination of single side SSPPs structure and curved ground mirror effect, the size of antenna is reduced effectively. The return loss simulation result of the antenna is shown in Fig. 5.29f, which shows that the antenna works in multi frequency band. The simulation results of the radiation pattern of the antenna are shown in Fig. 5.30. It has good omnidirectional performance in the horizontal direction, but the bandwidth is very narrow.

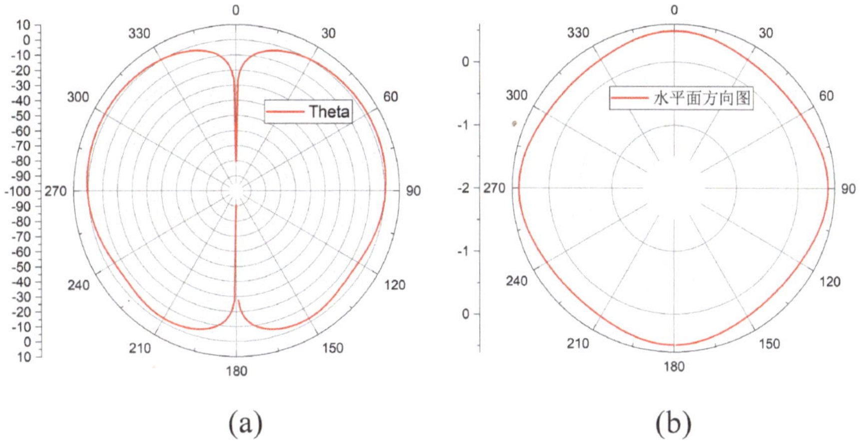

Fig. 5.30 Far-field patter of the omni-directional antenna based on single-side SSPPs with improved ground at 0.55 GHz. **a** E plane; **b** H plane

5.5 Summary of This Chapter

This chapter mainly uses SSPPs structure to design a miniaturized low profile broadband antenna in UHF band. The simulation results show that the SSPPs antenna has the characteristics of miniaturization, broadband, omnidirectional and vertical polarization in UHF band. This kind of antenna is very suitable for integration and application with modern wireless communication system.

References

1. H.F. Ma, X.P. Shen, Q. Cheng, W.X. Jiang, T.J. Cui, Broadband and high-efficiency conversion from guided waves to spoof surface plasmon polaritons. Laser Photon. Rev. **8**(1), 146–151 (2014)
2. D. Wang et al., A high-efficiency broadband omnidirectional UHF patch antenna applying surface plasmon polaritons for handheld terminals. IEEE Ant. Wirel. Propag. Lett. **17**(2), 283–286 (2018)
3. S. Quan, J. Zhou, H. Wang, Y. Huang, in *Dual Band Circular Polarized Low-Profile Air-Supported Folded Patch Antenna at UHF Band*. Proceedings of the 2015 IEEE 4th Asia-Pacific Conference on Antennas and Propagation, Kuta, Indonesia, 2015, pp. 197–198
4. G.S. Karthikeya, N. Agnihotri, in *A Conformal UHF Antenna for Cargo Helicopter Belly*. Proceedings of the 2015 IEEE 5th Asia-Pacific Conference on Antennas and Propagation, 2016, pp. 285–286
5. W. Wu, B. Guan, in *A Low-Profile Broadband UHF Base Station Antenna Based on Lumped-Element Matching Network*. Proceedings of the IEEE International Symposium on Antennas, Propagation and EM Theory, 2017, pp. 9–12
6. J. Wu, K. Sarabandi, Reactive impedance surface TM mode slow wave for patch antenna miniaturization [AMTAcorner]. IEEE Ant. Propag. Mag. **56**(6), 279–293 (2014)

7. S.L. Zuo, Y.Z. Yin, Z.Y. Zhang, K. Song, Enhanced bandwidth of low-profile sleeve monopole antenna for indoor base station application. Electron. Lett. **46**(24), 1587–1588 (2010)
8. J. Oh, K. Sarabandi, in *A Low-Profile Omnidirectional Planar Antenna with Vertical Polarization Employing Two In-Phase Elements*. 2011 XXXth URSI General Assembly and Scientific Symposium, Istanbul, 2011, pp. 1–4
9. L. Economou, R.J. Langley, Patch antenna equivalent to simple monopole. Electron. Lett. **33** (9), 727–729 (1997)
10. L. Yang, Z. Xing, S. Xu, D. Zhang, Y. Li, Design of a 3D top-loaded low-profile sleeve antenna for UAV applications. IEEE Access **8**, 64718–64724 (2020)
11. Q. Hou, H. Tang, Y. Liu, X. Zhao, Dual-frequency and broadband circular patch antennas with a monopole-type pattern based on epsilon-negative transmission line. IEEE Ant. Wirel. Propag. Lett. **11**, 442–445 (2012)
12. J. Liu, Q. Xue, H. Wong, H.W. Lai, Y. Long, Design and analysis of a low-profile and broadband microstrip monopolar patch antenna. IEEE Trans. Ant. Propag. **61**(1), 11–18 (2013)
13. A. Elsherbini, K. Sarabandi, Very low-profile top-loaded UWB coupled sectorial loops antenna. IEEE Ant. Wirel. Propag. Lett. **10**, 800–803 (2011)
14. L. Wang, B.L.G. Jonsson, J.R. Mosig, in *Wideband Omnidirectional Planar Antenna with Vertical Polarization*. 2017 11th European Conference on Antennas and Propagation (EUCAP), Paris, 2017, pp. 3865–3868
15. J.B. Pendry, L. Martín-Moreno, F.J. Garcia-Vidal, Mimicking surface plasmons with structured surfaces. Science **305**(5685), 847–848 (2004)
16. J. Han et al., in *A Low-Profile Dual-Band Antenna with Vertically Polarized Omnidirectional Radiation*. 2020 IEEE Asia-Pacific Microwave Conference (APMC), Hong Kong, Hong Kong, 2020, pp. 206–207
17. A.P. Hibbins, B.R. Evans, J.R. Sambles, Experimental verification of designer surface plasmons. Science **308**(5722), 670–672 (2005)
18. K.J. Zhuang, J.P. Geng, K. Wang, Pattern reconfigurable antenna applying spoof surface plasmon polaritons for wide angle beam steering. IEEE Access **7**, 15444–15451 (2019)
19. D. Wang, J.P. Geng, in *A Surface Plasmon Polariton Inspired Patch Antenna*. APS 2017
20. K. Wang, J.P. Geng, H. Zhou, C.H. Ren, R.H. Jin, X.L. Liang, W.R. Zhu, in *Pattern Reconfigurable Antenna Applying Spoof Surface Plasmon Polaritons*. EUCAP 2019
21. C. Xie et al., in *A High-Gain Broadband Omnidirectional Antenna Applying Spoof Surface Plasmon Polaritons with High Bearing Strength*. 2019 International Workshop on Electromagnetics: Applications and Student Innovation Competition (iWEM), Qingdao, China, 2018, pp. 1–2

Chapter 6
Study on the Rotated Spoof Surface Plasmon Polaritons Structure

Junping Geng, Chao Xie, and Weinan Gao

Spoof surface plasmon polaritons (SSPPs) structure has shown many novel electromagnetic performances, such as running wave in SSPPs structure with low loss, dispersive characteristics and strongly binding the electromagnetic wave in SSPPs structure, which are often used as transmission wave guide, diveiders and band filters in microwave RF devices. SSPPs structure can support the slow wave of surface wave mode, and it is difficult to directly radiate the bound EM energy into free space. Recently, periodic modulation of SSPPs structures are applied to increase the radiation efficiency of the SSPPs antenna. In the early days, the antenna based on SSPPs structure are mainly end-fire antenna (odd mode) and broadside radiation antenna (even mode). Later, we want to perform both the even mode and odd mode, and proposed the pattern reconfigurable SSPPs antenna to combine these two modes [1, 2]. The low profile, low cost and easy fabrication are the advantages of the existing SSPPs antenna. But the SSPPs antenna always shows a single polarization, and the even mode radiation pattern is not circular. To the problems of polarization and roundness of the far field pattern of SSPPs antenna, the rotating spoof surface plasmon polaritons structure (R-SSPPs) is presented to improve the SSPPs antennas in this chapter. By exciting the R-SSPPs structure with different feed methods, the end-fire CP SSPPs antenna and the broadside radiation SSPPs antenna with omnidirectional pattern are investigated.

J. Geng (✉) · C. Xie · W. Gao
Department of Electronic Engineering, Shanghai Jiao Tong University, Shanghai 200240, China
e-mail: gengjunp@sjtu.edu.cn

J. Geng et al., *Spoof Surface Plasmon Polaritons Antenna*, Modern Antenna,
https://doi.org/10.1007/978-981-16-4721-5_6

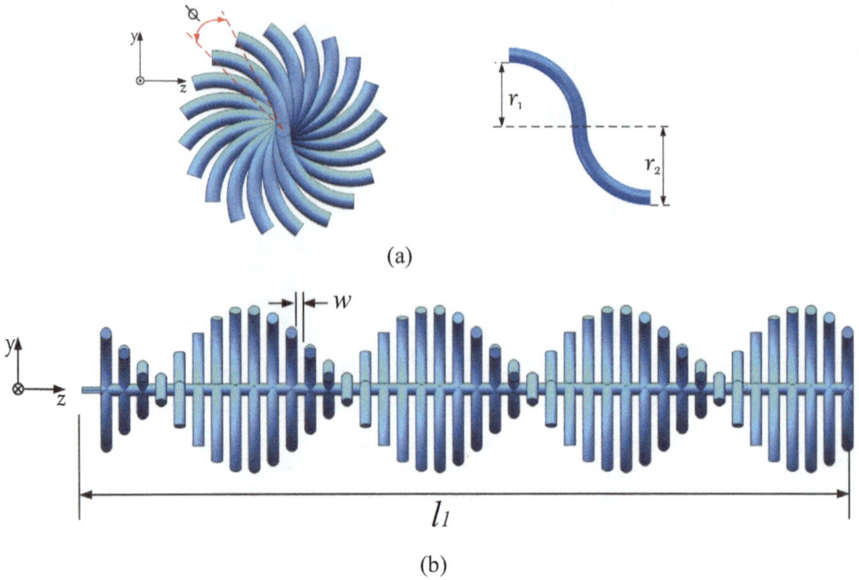

Fig. 6.1 Geometry of the rotated SSPPs structure. **a** Lateral view of the structure and the S-shaped structure; **b** front view. Figure reproduced with permission from Ref. [3], © 2021 IEEE

6.1 Study on the Characteristics of Rotated Spoof Surface Plasmon Polaritons Structure

Figure 6.1 is a schematic diagram of the structure of the rotating spoof surface plasmon polaritons (R-SSPPs) [3]. N S-shaped elements are arranged in parallel with equal spacing w along the axis z to form the R-SSPPs structure. The adjacent S-shaped arms rotate coaxially with angle step φ on the orthogonal planes which are parallel to xoy plane, and the length of the S-shaped arm is $\frac{r_1+r_2}{2}\pi$. Specifically, the R-SSPPs structure proposed in this chapter is obtained through three stages, as shown in Fig. 6.2. The first step is to construct the traditional cylindrical double arm SSPPs structure, as shown in Fig. 6.1a, which consists of N double arm units arranged in parallel along the z axis with equal spacing w. the double side arm units are cylindrical with a height of $l_e \approx \frac{r_1+r_2}{2}\pi$. The traditional SSPPs structure is named Str#1. Secondly, on the basis of Str#1, in the parallel plane perpendicular to the axis of SSPPs, we rotate each cylindrical arm coaxially at the angle of φ in turn, so as to obtain a new SSPPs structure, which is called Str#2, as shown in Fig. 6.2b. Thirdly, the double side SSPPs arm is bent into S-shaped, and the SSPPs unit in Str#2 is replaced by the S-shaped arm to obtain the third SSPPs structure, which is called the rotating SSPPs structure (R-SSPPs), just proposed in this chapter, which is called Str#3, as shown in Fig. 6.2c.

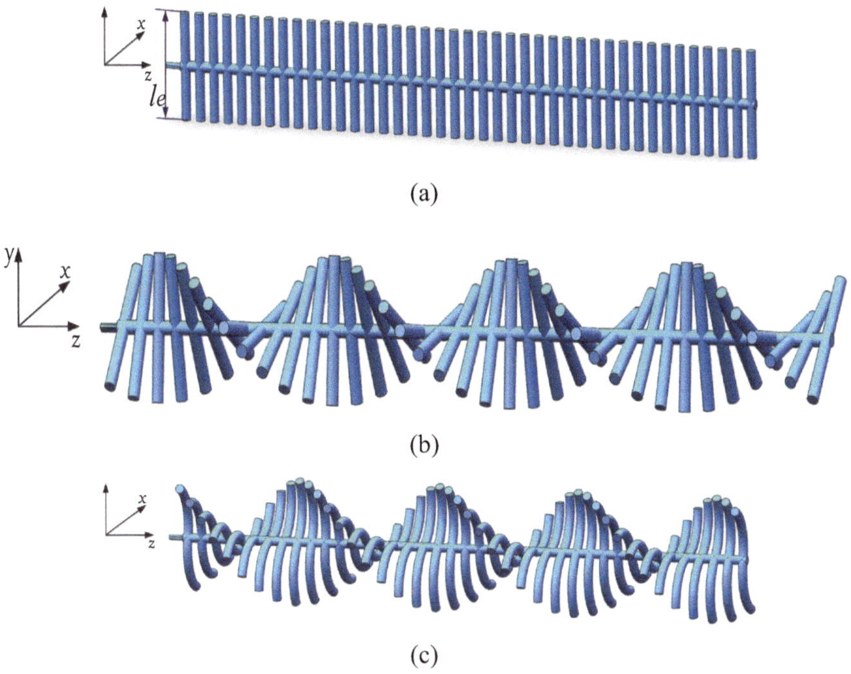

Fig. 6.2 Three stages of evolution of the proposed SSPPs structure: **a** Str#1; **b** Str#2; **c** Str#3. Figure reproduced with permission from Ref. [3], © 2021 IEEE

Firstly, the dispersion characteristics of three kinds of SSPPs structures are analyzed by simulation. The main parameters are $N = 40$, $w = 2$ mm, $r_1 = 11.75$ mm, $r_2 = 14.75$ mm, $l_e = 21$ mm and $l_1 = 208$ mm.

The dispersion characteristics of three SSPPs Str#1, Str#2 and Str#3 are shown in Fig. 6.3. Obviously, the dispersion curves of the three SSPPs deviate from the propagation curve of light wave and distribute on the right side of the light, which also shows that the wave vectors of Str#1, Str#2 and Str#3 SSPPs are larger than that of light wave, indicating that the three SSPPs strctures can transmit slow waves and all are SSPPs modes. In addition, it can be seen that the dispersion curves of Str#1 and Str#2 of SSPPs structure are basically same, but only the dispersion curves of Str#3 is below the dispersion curves of Str#1 and Str#2. The characteristics of these curves show that Str#3 of SSPPs structure has stronger electric field binding ability and stronger SSPPs performance in the frequency range of 2–3 GHz.

In addition, the plane wave is also introduced to excite the SSPPs structure. Different excitation methods will excite different SSPPs modes: odd mode and even mode. Besides the traditional SSPPs structure Str#1 can support these two kinds of slow wave modes, similar slow wave modes will be transmited in SSPPs structures Str#2 and Str#3 too. In the following part, the radiation principle of the three SSPPs structures is explored through the scattering cross section (RCS), electric field

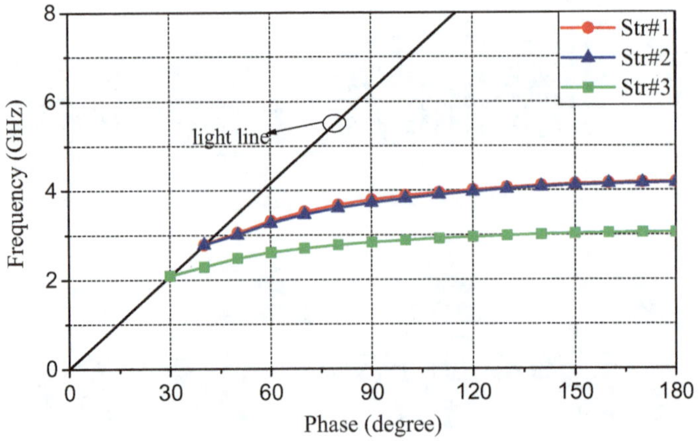

Fig. 6.3 The dispersion curves of the Str#1, Str#2 and Str#3. Figure reproduced with permission from Ref. [3], © 2021 IEEE

distribution, power flow and far-field mode when the three SSPPs structures are excited by plane wave.

6.1.1 Odd Mode of Rotated Spoof Surface Plasmon Polaritons Structure

As shown in Fig. 6.4a, when the traditional SSPPs structure, Str#1 is excited by the planewave along z axis, it will work in odd mode, the current distribution on its arm just likes small dipole, as shown in Fig. 6.5a. Here, the maximum current magnitude is set as $\mathbf{J}_{\mathrm{m}}$, and the attenuation of current amplitude along the axis is ignored, and the current distribution on the nth arm ($z = nw$) of the SSPPs Str#1 can be expressed as

$$\vec{\mathrm{J}}(z,t) = \vec{\mathrm{a}}_y \mathrm{J}_m \sin(kr_0 - k|y|)\cos(\omega t - knw) \tag{6.1}$$

Equation (6.1) is a typical linear polarization (LP) current distribution.

Similarly, for the rotated SSPPs structure Str#2, the induced current distribution vector on the nth ($z = nw$, $\boldsymbol{r}$) arm can be devided into x-component and y-component.

Here, $\boldsymbol{r}$ is the position vecroe, and can be written as

$$\boldsymbol{r} = \vec{\mathrm{a}}_x x + \vec{\mathrm{a}}_y y = \vec{\mathrm{a}}_x r\sin(nd\varphi) + \vec{\mathrm{a}}_y r\cos(nd\varphi) \tag{6.2}$$

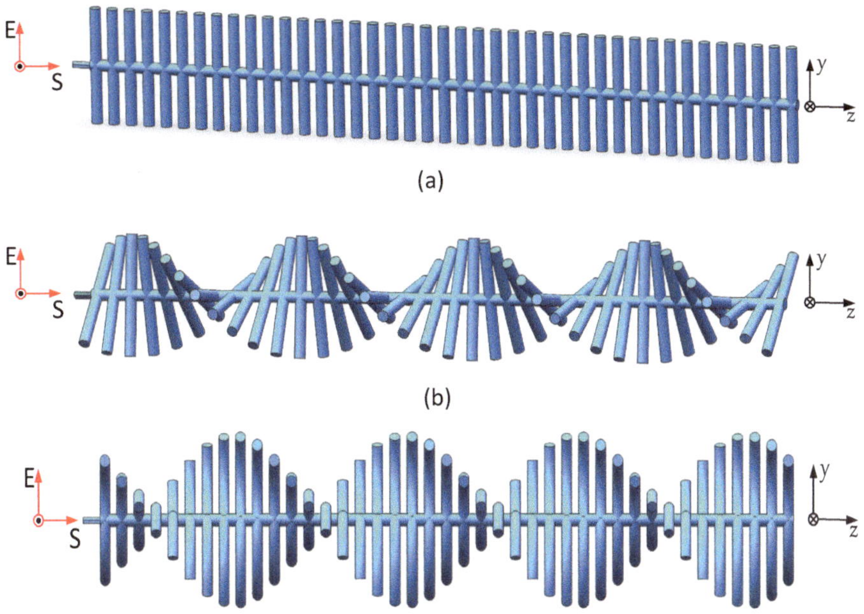

Fig. 6.4 The proposed SSPPs structures is excited by plane wave to work in odd mode. Figure reproduced with permission from Ref. [3], © 2021 IEEE

And the current distribution is just written as

$$\begin{aligned} \vec{\mathrm{J}}(z = nw, \boldsymbol{r}) =& \vec{\mathrm{a}}_y \mathrm{J}_m \sin(kr_0 - k|r|) \cos(nd\varphi) \cos(\omega t - knw) \\ & - \vec{\mathrm{a}}_x \mathrm{J}_m \sin(kr_0 - k|r|) \sin(nd\varphi) \cos(\omega t - knw) \\ =& \vec{\mathrm{a}}_y \mathrm{J}_m \sin(kr_0 - k|r|) \cos(nd\varphi) \cos(\omega t - knw) \\ & + \vec{\mathrm{a}}_x \mathrm{J}_m \sin(kr_0 - k|r|) \cos(nd\varphi + 90^\circ) \cos(\omega t - knw) \\ =& \mathrm{J}_m \sin(kr_0 - k|r|) \cos(\omega t - knw) \left[\vec{\mathrm{a}}_x \cos(nd\varphi + 90^\circ) + \vec{\mathrm{a}}_y \cos(nd\varphi) \right] \end{aligned} \tag{6.3}$$

It is noted that current distribution Eq. (6.3) is the time harmonic variation function of time and space at the given frequency, and it rotates in space, which varies like right hand circle polarization (RHCP) along z axis of SSPPs structure, and the schematic diagram of the currents distributions on the Str#2 are given in Fig. 6.5b.

To the rotated SSPPs structure Str#3, the variable arc length between the point on the arm to the original point is assumed as s, and the length of one side of the s-shaped arm is $s_0 = r_0$. Then, the current distribution vector on the nth ($z = nw$, s) s-shaped arm can be devided into x-component and y-component too.

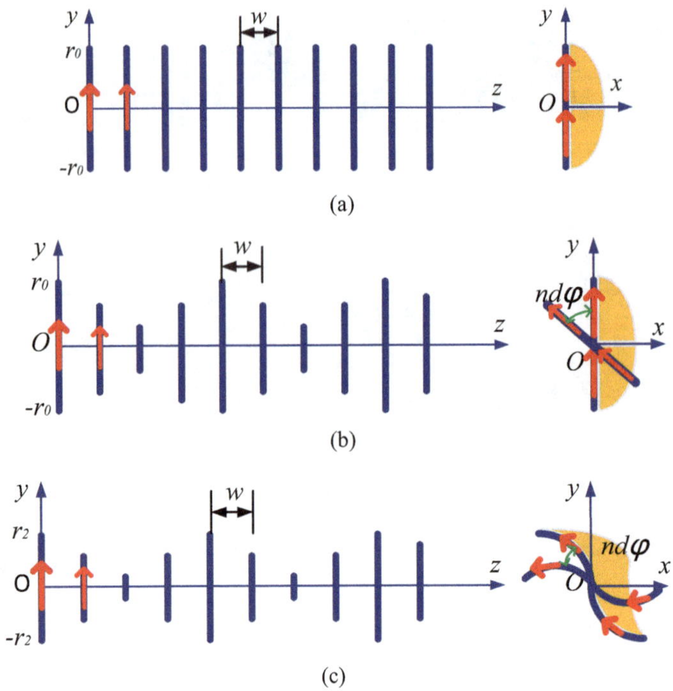

Fig. 6.5 Current distribution diagram in the SSPPs structure. **a** Str#1; **b** Str#2 and **c** Str#3 in the odd mode. Figure reproduced with permission from Ref. [3], © 2021 IEEE

$$s = r_0 \arcsin\left(\frac{\pi y}{2r_0}\right) \tag{6.4}$$

$$\vec{a}_s = \vec{a}_x \frac{\pi x}{2r_0} + \vec{a}_y \frac{\pi y}{2r_0} \quad (z = 0) \tag{6.5}$$

And the current distribution on the nth ($z = nw$, s) s-shaped arm can be written as

$$\begin{aligned} \vec{J}(z = 0, s) &= \vec{a}_s J_m \sin(kr_0 - k|s|)\cos(\omega t) \\ &= \vec{a}_s J_m \sin(kr_0 - k|s|)\cos(\omega t) \\ &= \left(\vec{a}_x \frac{\pi x}{2r_0} + \vec{a}_y \frac{\pi y}{2r_0}\right) J_m \sin(kr_0 - k|s|)\cos(\omega t) \end{aligned} \tag{6.6}$$

$$\begin{aligned}\vec{a}_s(z = nw, s) &= \vec{a}_x \frac{\pi x}{2r_0}\cos(nd\varphi) + \vec{a}_x \frac{\pi y}{2r_0}\sin(nd\varphi) - \vec{a}_y \frac{\pi x}{2r_0}\sin(nd\varphi) + \vec{a}_y \frac{\pi y}{2r_0} * \cos(nd\varphi) \\ &= \vec{a}_x\left(\frac{\pi x}{2r_0}\cos(nd\varphi) + \frac{\pi y}{2r_0}\cos(nd\varphi - 90^\circ)\right) \\ &\quad - \vec{a}_y\left(\frac{\pi x}{2r_0}\cos(nd\varphi - 90^\circ) + \frac{\pi y}{2r_0} * \cos(nd\varphi)\right)\end{aligned} \tag{6.7}$$

$$\begin{aligned}\vec{J}(z = nw, s) &= \vec{a}_y J_m \sin(kr_0 - k|s|)\sin(nd\varphi)\cos(\omega t - knw) \\ &\quad + \vec{a}_x J_m \sin(kr_0 - k|s|)\cos(nd\varphi)\cos(\omega t - knw) \\ &= \vec{a}_y J_m \sin(kr_0 - k|s|)\cos(nd\varphi - 90^\circ)\cos(\omega t - knw) \\ &\quad + \vec{a}_x J_m \sin(kr_0 - k|s|)\cos(nd\varphi)\cos(\omega t - knw) \\ &= J_m \sin(kr_0 - k|s|)\cos(\omega t - knw)\left[\vec{a}_x \cos(nd\varphi) + \vec{a}_y \cos(nd\varphi - 90^\circ)\right]\end{aligned} \tag{6.8}$$

It is noted that current distribution Eq. (6.8) is the time harmonic variation function of time and space at the given frequency, and it rotates in space along the z axis, which transimts like right hand circle polarization (RHCP) along z axis of SSPPs structure too. And the schematic diagram of the currents distributions on the Str#3 are given in Fig. 6.5c.

As shown in Fig. 6.4, a linearly polarized plane electromagnetic wave is used to irradiate the SSPPs structure from the left side to observe the scattering characteristics of the others. The scattering curves of the three SSPPs structures (Str#1, Str#2 and Str#3) under the irradiation of the linearly polarized plane wave are shown in Fig. 6.6. It is obvious that the peak values of the three RCS curves are all around 2.45 GHz, which indicates that the resonance points of Str#1, Str#2 and Str#3 are all around 2.45 GHz. Therefore, $f = 2.45$ GHz is selected as the

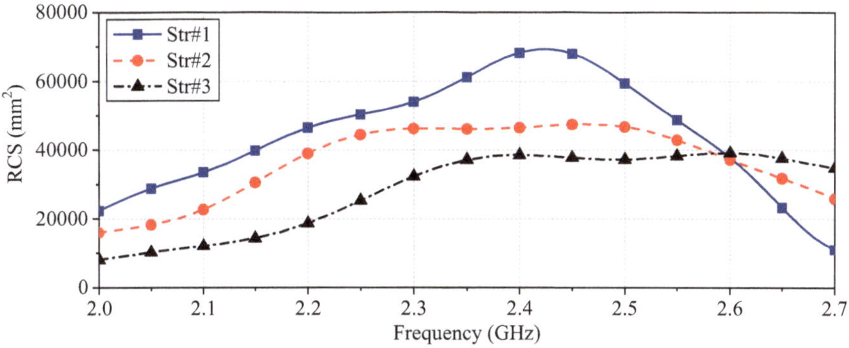

Fig. 6.6 RCS curves of the Str#1, Str#2 and Str#3 (in the odd mode) under plane wave in Fig. 6.4a. Figure reproduced with permission from Ref. [3], © 2021 IEEE

observation point, and the electric field distributions, power flows and far-field modes of these three SSPPs structures are analyzed and compared at this frequency.

The electric field distributions of three SSPPs Str#1, Str#2 and Str#3 at 2.45 GHz are shown in Fig. 6.7. It can be seen that the electric field distributions of the three SSPPs structures are periodic and anti symmetric in the YOZ plane. According to the working principle and mode definition of SSPPs structure, it is obvious that the three SSPPs structures, Str#1, Str#2 and Str#3 are excited by linearly polarized plane waves. Figure 6.7c also shows the electric field distribution at different times (t = 0, T/4, T/2, 3 T/4) on the D–D section of Str#3. The direction of the main tangential electric field on the D–D cross section rotates counter-clockwise with time, which means that the Str#3 structure may transform the linearly polarized wave into the circularly polarized wave.

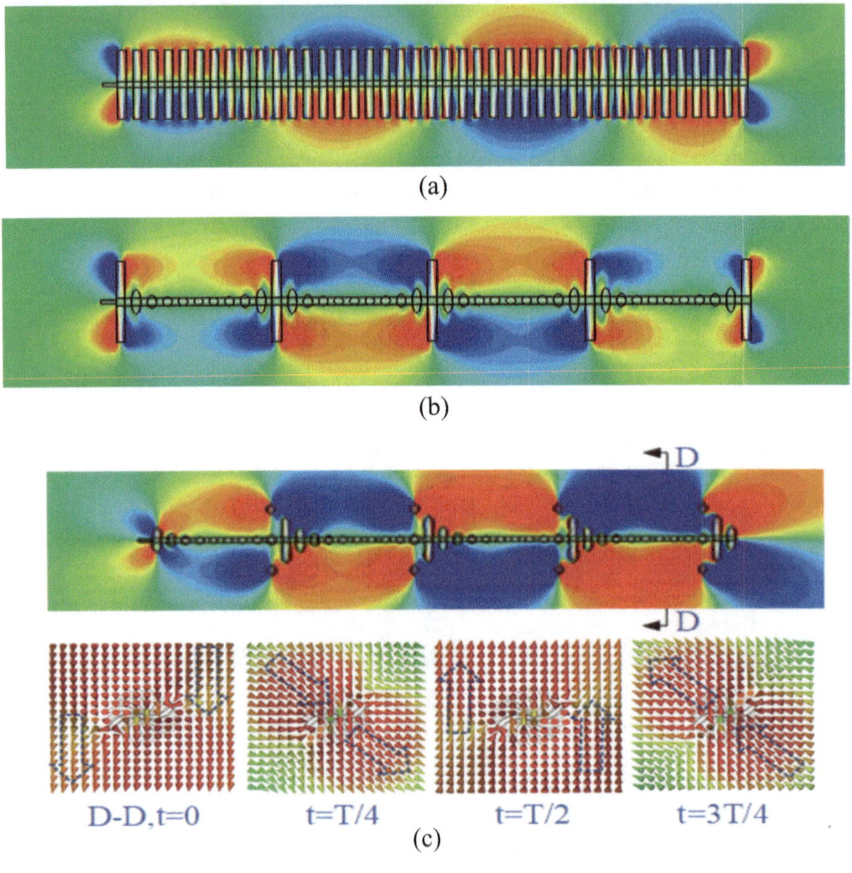

Fig. 6.7 The electric filed distribution of the **a** Str#1; **b** Str#2 and **c** Str#3 (in the odd mode) under plane wave. In the cut plane D–D in Str#3, the direction of the main tangent electric field rotates counter clockwise with time varying in one period T. Figure reproduced with permission from Ref. [3], © 2021 IEEE

The power flow vector distributions of these three SSPPs at $f = 2.45$ GHz are shown in Fig. 6.8a, c and e. It can be seen that under the irradiation of linear polarized plane wave, the electromagnetic energy is mainly confined in the SSPPs structure and transmitted along the right end of the SSPPs structure axis. At the end of the structure, the electromagnetic energy is radiated to the free space, and the three SSPPs structures finally form the end fire pattern, as shown in Fig. 6.8b, d and f. And the 3 dB width of the main lobe of Str#3 is 55.4°–63.3° in Fig. 6.8f. In general, the rotated SSPPs structures can transmit the odd mode running wace similar as that of the traditional SSPPs structure, just like Str#1 [7].

6.1.2 Even Mode of Rotated Spoof Surface Plasmon Polaritons Structure

Similar as the above analysis, when the traditional SSPPs structure, Str#1, is excited by the planewave vertical to z axis, as shown in Fig. 6.9a, it will work in even

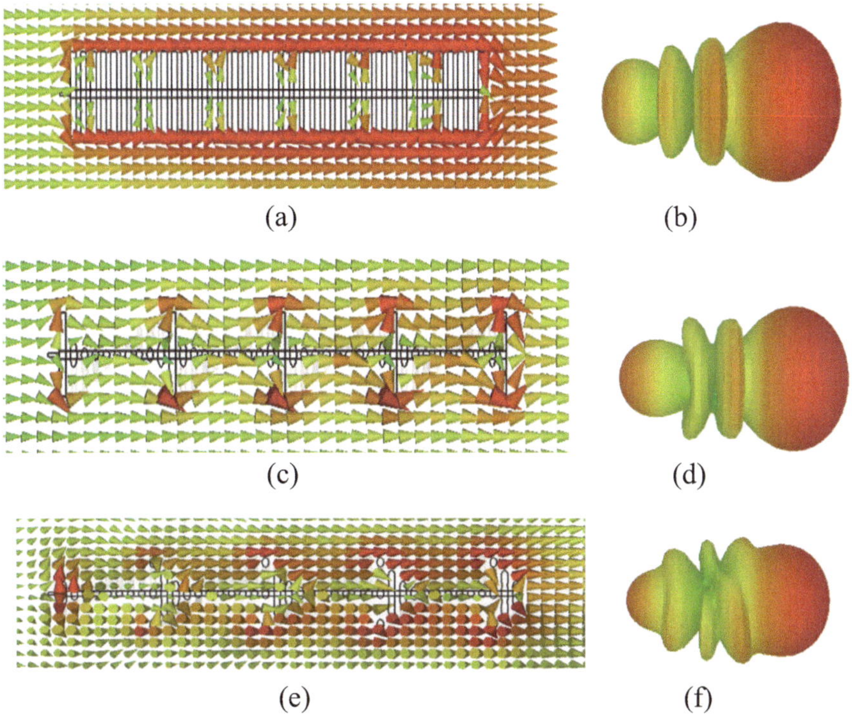

Fig. 6.8 The odd mode: the power flow (**a**) and radiation pattern (**b**) of the Str#1; the power flow (**c**) and radiation pattern (**d**) of the Str#2; the power flow (**e**) and radiation pattern (**f**) of the Str#3 at 2.45 GHz. Figure reproduced with permission from Ref. [3], © 2021 IEEE

mode, the current distribution on its upper arm just likes small dipole, as shown in Fig. 6.10a. And the current distribution is similar as the odd mode Eq. (6.1).

To the current distribution on the arm of the rotated SSPPs structure, there may appear opposite current directions, as shown Fig. 6.9b, c. In fact, for the asymmetrical structure, the opposite current distributions on the rotated SSPPs structure in Fig. 6.9b, c are not symmetrical, and are called quasi-even mode.

In Fig. 6.9b, to the rotated SSPPs structure Str#2 in quasi-even mode, the current distribution on the upper half arm at $z = nw$ can be expressed as

$$
\begin{aligned}
\vec{\mathrm{J}}_u(z = nw, r) = & -\vec{\mathrm{a}}_y \mathrm{J}_m \sin(kr_0 - k|r|)\cos(nd\varphi)\cos(\omega t - knw) \\
& + \vec{\mathrm{a}}_x \mathrm{J}_m \sin(kr_0 - k|r|)\sin(nd\varphi)\cos(\omega t - knw) \\
= & -\vec{\mathrm{a}}_y \mathrm{J}_m \sin(kr_0 - k|r|)\cos(nd\varphi)\cos(\omega t - knw) \\
& + \vec{\mathrm{a}}_x \mathrm{J}_m \sin(kr_0 - k|r|)\cos(nd\varphi - 90^\circ)\cos(\omega t - knw) \\
= & -\mathrm{J}_m \sin(kr_0 - k|r|)\cos(\omega t - knw)\left[\vec{\mathrm{a}}_x \cos(nd\varphi + 90^\circ) + \vec{\mathrm{a}}_y \cos(nd\varphi)\right]
\end{aligned}
\tag{6.9}
$$

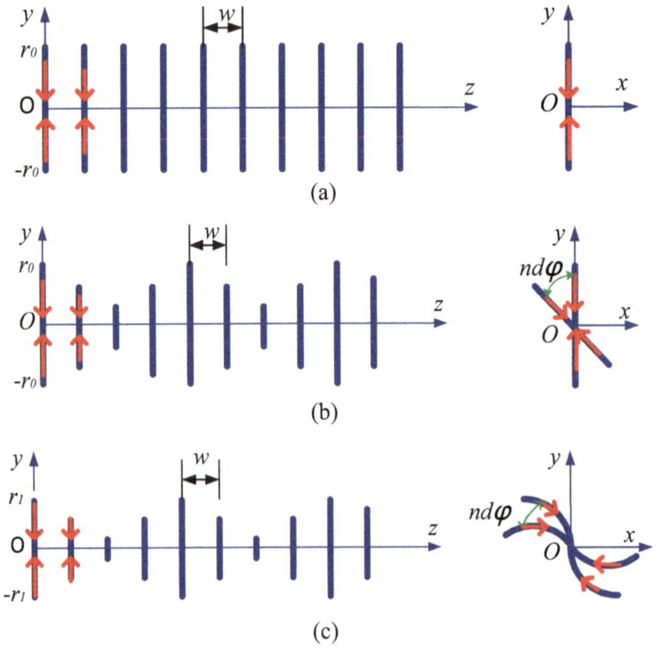

Fig. 6.9 Current distribution diagram in the SSPPs structure. **a** Str#1 in the even mode; **b** Str#2 and **c** Str#3 in the quasi-even mode. Figure reproduced with permission from Ref. [3], © 2021 IEEE

It is noted that $\vec{J}_u$ is just like left hand circular polarization distribution. To the lower half arm, the current distribution on at $z = nw$ can be similarly expressed as

$$\begin{aligned}\vec{J}_l(z = nw, r) =& \vec{a}_y J_m \sin(kr_0 - k|r|)\cos(nd\varphi)\cos(\omega t - knw) \\ &- \vec{a}_x J_m \sin(kr_0 - k|r|)\sin(nd\varphi)\cos(\omega t - knw) \\ =& \vec{a}_y J_m \sin(kr_0 - k|r|)\cos(nd\varphi)\cos(\omega t - knw) \\ &+ \vec{a}_x J_m \sin(kr_0 - k|r|)\cos(nd\varphi + 90^\circ)\cos(\omega t - knw) \\ =& J_m \sin(kr_0 - k|r|)\cos(\omega t - knw)\left[\vec{a}_x \cos(nd\varphi + 90^\circ) + \vec{a}_y \cos(nd\varphi)\right]\end{aligned} \tag{6.10}$$

It is clear that $\vec{J}_l$ is just like right hand circular polarization distribution. In the space, these two current parts, $\vec{J}_u$ and $\vec{J}_l$, radiated the opposite polarization fields and they maybe cancel each other, and results to linear polarization result. So, the current distribution on the rotated SSPPs structure Str#2 should be the dominant linear polarization in the $nd\varphi$ direction at $z = nw$.

Similar to the rotated SSPPs structure Str#3 in quasi-even mode, in Fig. 6.9c, the current distribution on the upper half arm at $z = nw$ can be expressed as.

$$\begin{aligned}\vec{J}_u(z = nw, s) =& -\vec{a}_s(z = nw) J_m \sin(kr_0 - k|s|)\cos(\omega t - nkw) \\ =& J_m \sin(kr_0 - k|s|)\cos(\omega t - nkw) * \\ &\left\{-\vec{a}_x\left[\frac{\pi x}{2r_0} * \cos(nd\varphi) + \frac{\pi y}{2r_0} * \cos(nd\varphi + 90^\circ)\right]\right. \\ &\left.+ \vec{a}_y\left[\frac{\pi x}{2r_0}\cos(nd\varphi - 90^\circ) + \frac{\pi y}{2r_0} * \cos(nd\varphi)\right]\right\} \\ =& -\left(\frac{\pi x}{2r_0}(\vec{a}_x * \cos(nd\varphi) + \vec{a}_y \cos(nd\varphi - 90^\circ)\right) \\ &+ \frac{\pi y}{2r_0}\left(\vec{a}_x * \cos(nd\varphi + 90^\circ) + \vec{a}_y \cos(nd\varphi)\right) * J_m \sin(kr_0 - k|s|)\cos(\omega t - nkw)\end{aligned} \tag{6.11}$$

$\vec{J}_u$ is just like the left hand circular polarization distribution. And the current distribution on the lower half arm at $z = nw$ can be similarly achieved as

$$
\begin{aligned}
\vec{J}_l(z = nw, s) &= \vec{a}_s(z = nw, s) J_m \sin(kr_0 - k|s|)\cos(\omega t - nkw) \\
&= \left\{ \vec{a}_x \left[\frac{\pi x}{2r_0} * \cos(nd\varphi) + \frac{\pi y}{2r_0} * \cos(nd\varphi + 90^\circ) \right] \right. \\
&\quad \left. + \vec{a}_y \left[\frac{\pi x}{2r_0} cos(nd\varphi - 90^\circ) + \frac{\pi y}{2r_0} * \cos(nd\varphi) \right] \right\} \\
&\quad * J_m \sin(kr_0 - k|s|)\cos(\omega t - nkw) \\
&= \left(\frac{\pi x}{2r_0} \vec{a}_x \cos(nd\varphi) + \vec{a}_y cos(nd\varphi - 90^\circ) \right) \\
&\quad + \frac{\pi y}{2r_0} \left(\vec{a}_x \cos(nd\varphi + 90^\circ) + \vec{a}_y \cos(nd\varphi) \right)) * J_m \sin(kr_0 - k|s|)\cos(\omega t - nkw)
\end{aligned}
\tag{6.12}
$$

$\vec{J}_l$ is just like right hand circular polarization distribution. In the space, these two current parts, $\vec{J}_u$ and $\vec{J}_l$, radiate the opposite polarization fields and they maybe cancel each other, and results to linear polarization result. So, the current distribution on the rotated SSPPs structure Str#3 should be the dominant linear polarization in the $(nd\varphi - 45^\circ)$ direction at $z = nw$.

Similar to the odd mode, these three SSPPs structure can also be excited to transmit the even mode or quasi even mode electromagnetic waves. As shown in Fig. 6.10, the propagation direction of the plane wave is perpendicular to the z-axis. Figure 6.11 shows the RCS curves of these three SSPPs structures under the side irradiation of the linearly polarized plane wave. It can be seen from the curves that the resonance frequencies of Str#1, Str#2 and Str#3 are not the same, and they are 2.4 GHz @Str#1, 2.31 GHz @Str#2 and 2.37 GHz @Str#3. The distance between the corresponding points on the adjacent rotating arms of Str#2 is greater than that of Str#1, so the resonance frequency of Str#2 moves to lower frequency. On the other hand, the length of s-arm in Str#3 is the same as that of Str#2, so the equivalent radial length of s-arm is less than that of Str#2, and the resonance frequency of Str#3 is higher than that of Str#2.

Referring to the maximum point of RCS curves in Fig. 6.11, we select $f = 2.4$ GHz as the even mode/quasi even mode observation frequency of SSPPs structure Str#1 and Str#3. And then mainly analyze the electric field distributions, power flow distributions and far-field patterns of the two SSPPs structures at this frequency. The observation point of SSPPs structure Str#2 is at $f = 2.33$ GHz. In Fig. 6.12, the electric field distributions of these three SSPPs structures are compared at the observation frequencies. From the simulation results of the electric field distribution, it can be clearly seen that the electric field distributions of these three SSPPs strutures in the *yoz* plane are symmetrical, which is very similar to that of the traditional even mode SSPPs wave. In Fig. 6.12c, four cross sections A–A, B–B, C–C and D–D are selected on Str#3, and the distance between them is about half wavelength. It can be seen from the electric field vector distribution on these cross sections that the electric field vector directions on these cross sections are radial,

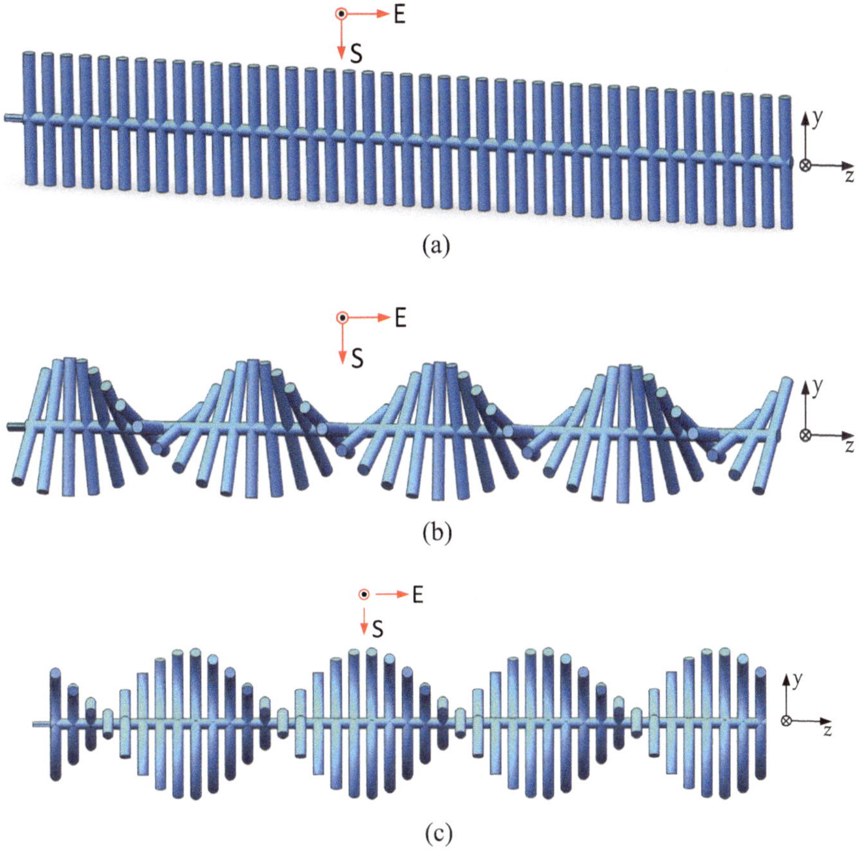

Fig. 6.10 The proposed SSPPs structure is excited by plane wave to work in even mode or quasi-even mode. **a** Traditional SSPPs structure, Str#1, in even mode; **b** rotated SSPPs Structure Str#2 in quasi-even mode and **c** rotated SSPPs Structure Str#3 in quasi-even. Figure reproduced with permission from Ref. [3], © 2021 IEEE

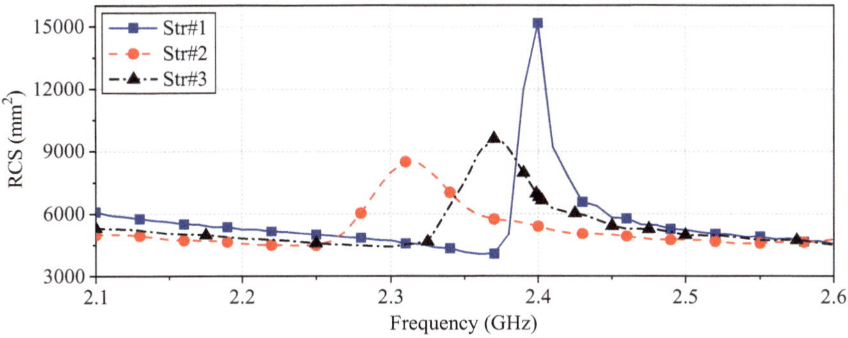

Fig. 6.11 RCS curves of the Str#1, Str#2 and Str#3 (in the even mode) under plane wave in Fig. 6.4b, the peaks appear at 2.4 GHz @Str#1, 2.31 GHz @Str#2, and 2.37 GHz @Str#3. Figure reproduced with permission from Ref. [3], © 2021 IEEE

and the electric field is almost uniformly distributed along the circumferential direction, which is very similar to the electric field distribution of the multipole. This kind of distribution is very conducive to the omni-directional radiation of these three SSPPs structure of Str#1, Str#2 and Str#3 in the *xoy* plane. Moreover, the direction of the electric field on these 4 cross-sections changes 180 degrees in space in turn, which indirectly indicates that the electromagnetic wave is propagating along the z-axis.

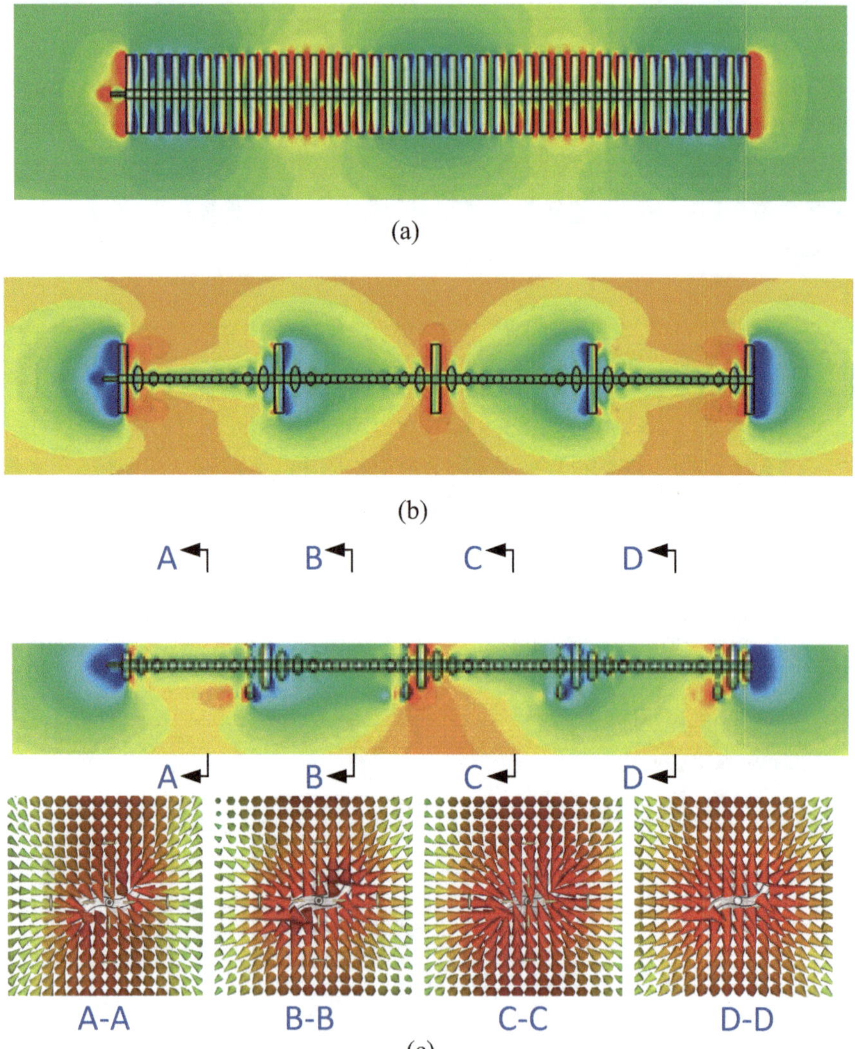

Fig. 6.12 The electric filed distribution of the Str#1, Str#2 and Str#3 (in the even mode) under plane wave. Figure reproduced with permission from Ref. [3], © 2021 IEEE

The power flow vector distributions of these three SSPPs Str#1, Str#2 and Str#3 at the observation frequencies are shown in Fig. 6.13a, c and e. It can be seen that the electromagnetic energy of these three SSPPs structures radiates mainly from the side of the structure axis to the free space under the side irradiation of the linear polarized plane wave, thus forming the omni-directional far-field pattern of the side direction in *xoy* plane, corresponding to Fig. 6.13b, d and f respectively. In the whole space, the main lobe 3 dB beamwidths of these three SSPPs structures are 19.1°–19.4°, 23.9°–30° and 28°–30° respectively.

From the above results and analysis, it can be seen that the R-SSPPs structure (Str#2 and Str#3) can support odd mode and quasi even mode SSPPs wave. When the SSPPs structure Str#3 works in odd mode, it can radiate circularly polarized end fire pattern by generating anti clockwise rotating tangential electric field; when the R-SSPPs structure Str#3 works in even mode, its beam width of side radiation pattern is narrow, but omnidirectional in *xoy* plane.

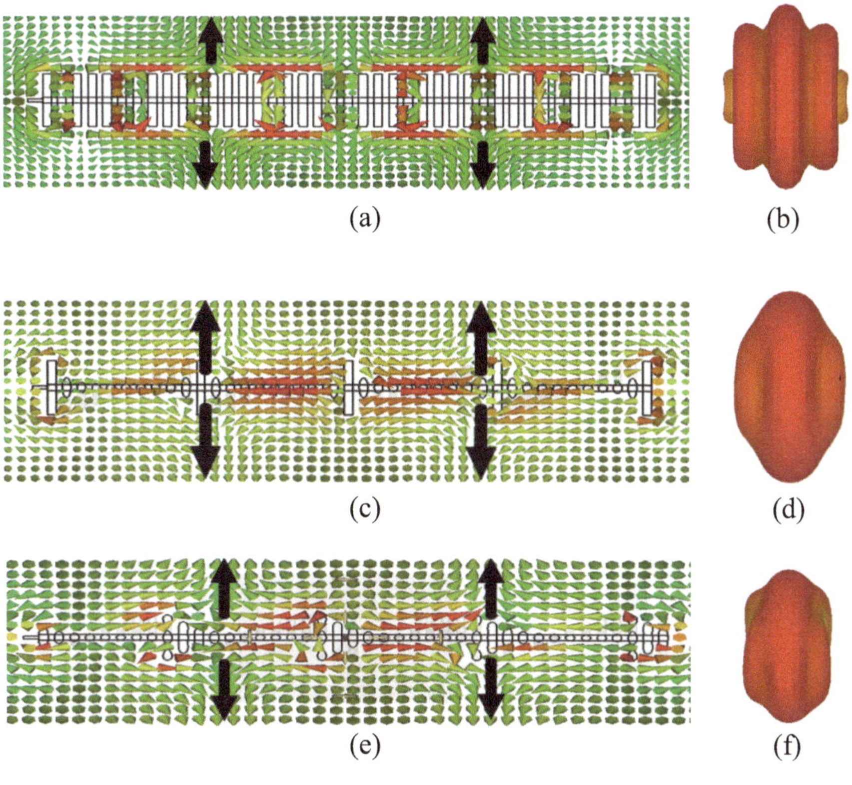

Fig. 6.13 The even mode: the power flow (**a**) and radiation pattern with 3 dB width 19.1°–19.4° (**b**) of the Str#1 at 2.4 GHz; the power flow (**c**) and radiation pattern with 3 dB width 23.9°–30° (**d**) of the Str#2 at 2.33 GHz; the power flow (**e**) and radiation patter. Figure reproduced with permission from Ref. [3], © 2021 IEEE

6.2 End-Fire Antenna Based on Rotated Spoof Surface Plasmon Polaritons

In the previous section, the transmission and radiation characteristics of R-SSPPs structure (Str#3) in odd/even modes are analyzed. Here, the actual feed structure is introduced to replace the linearly polarized plane wave in Fig. 6.14, as shown in Fig. 6.14. The details are discussed in detail below.

6.2.1 *Antenna Model and Simulation Verification*

The front view and feed structure details of R-SSPPs (Str#3) end fire antenna are shown in Fig. 6.14, which is called E-Ant.3. SMA connector is shown in gray colour in the figure, printed metal is blue colour, and substrate medium is orange colour. The end fire antenna mainly includes R-SSPPs transmission and radiation structure and monopole feed structure. The monopole is used to replace the linearly polarized plane wave in Fig. 6.14 to excite the odd mode SSPPs wave in the R-SSPPs structure. The monopole is placed perpendicular to the principal axis of the R-SSPPs structure and separated by a certain distance. In order to widen the bandwidth, the monopole is trapezoidal and printed on one side of the FR-4 dielectric plate. The thickness of the FR-4 dielectric plate is 1 mm, and there is a rectangular metal ground on the other side of the dielectric plate. The dielectric constant of the FR-4 plate is 4.4 and the loss tangent is 0.025. The main parameters of R-SSPPs structure_Str#3 are r_1 = 16 mm and r_2 = 13 mm.

The simulated electric field distribution of the antenna based on R-SSPPs structure Str#3 (E-Ant.3) at 2.45 GHz is shown in Fig. 6.15a. It can be seen that the

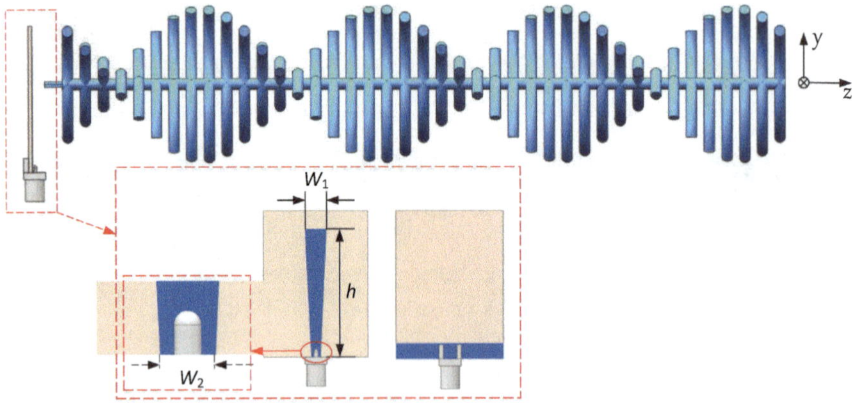

Fig. 6.14 The front view of the end-fire antenna based on Str#3 and the details of the feed. Figure reproduced with permission from Ref. [3], © 2021 IEEE

blue electric field and the red electric field alternately appear on one side, and the local anti symmetry appears. This kind of electric field distribution is mainly caused by the spatial rotation characteristics of Str#3 structure, and only partially shows the dipole electric field characteristics in the *yoz* plane. As shown in Fig. 6.15b, the dipole characteristic electric field distribution is still obvious in the *xoy* plane, which also indicates that the electric field of Str#3 rotates along the *z* axis in space, and makes the endfire antenna based on R-SSPPs structure have circular polarization characteristics.

The simulation diagram of power flow distribution of the antenna based on R-SSPPs structure Str#3 (E-Ant.3) at 2.45 GHz is shown in Fig. 6.15c. Obviously, most of the electromagnetic energy radiated from the printed monopole is coupled into the R-SSPPs structure, bound to the inside and surface of the R-SSPPs structure. The EM energy is transmitted to the right end along the main axis of the R-SSPPs structure, and radiated out at the end of the R-SSPPs structure to form end fire radiation pattern. The 3-D far-field pattern of the E-Ant.3 at 2.45 GHz is shown in Fig. 6.15d. It is obvious that the energy of the antenna is mainly distributed in the main lobe to the right along the main axis, and the rear lobe is very small, which is consistent with the power flow distribution in Fig. 6.15c.

For the sake of comparison, Str#1 and Str#2 are also excited and fed from the left side by the trapezoidal printed monopole respectively, which are called E-Ant.1 and E-Ant.2. And the arm lengths of Str#1 and Str#2 are $l_e = 45.5$ mm. The simulation results of return loss of the end fire antenna based on these three kinds of SSPPs structure are shown in Fig. 6.16a. Obviously, the impedance bandwidth ($|S_{11}| < -10$ dB) of the odd mode endfire antenna based on R-SSPPs Str#3 (E-Ant.3) can cover 2.2–2.65 GHz, which is larger than that of the end fire antenna based on Str#1 (E-Ant.1)and Str#2 (E-Ant.2). Figure 6.16b shows the axial ratio curve of the main radiation direction of the end fire antenna based on the three SSPPs structures. The simulated axial ratio bandwidth (AR < 3 dB) of E-Ant.2 is 152 MHz (2.353–2.505 MHz), and that of E-Ant.3 is 163 MHz (2.392–2.555 GHz). However, the end fire antenna based on the traditional dual arm SSPPs structure Str#1 (E-Ant.1) has a large axial ratio, which obviously does not have circular polarization characteristics.

The simulated gain curves of the end fire antennas based on these three SSPPs structure are plotted in Fig. 6.16b too. The gain of E-Ant.1is 5 $\sim$ 10 dBi in the available impedance band. And the gain in the CP band of the E-Ant.2 and E-Ant.3 are 6.86 $\sim$ 8.7 dBi and 6.77 $\sim$ 8 dBi, respectively.

In general, the end-fire antenna based on Str#3 (E-Ant.3) performs good AR bandwidth as well as the end-fire antenna based on Str#2 (E-Ant.2), and its impedance bandwidth is better but with a smaller aperture than E-Ant.2.

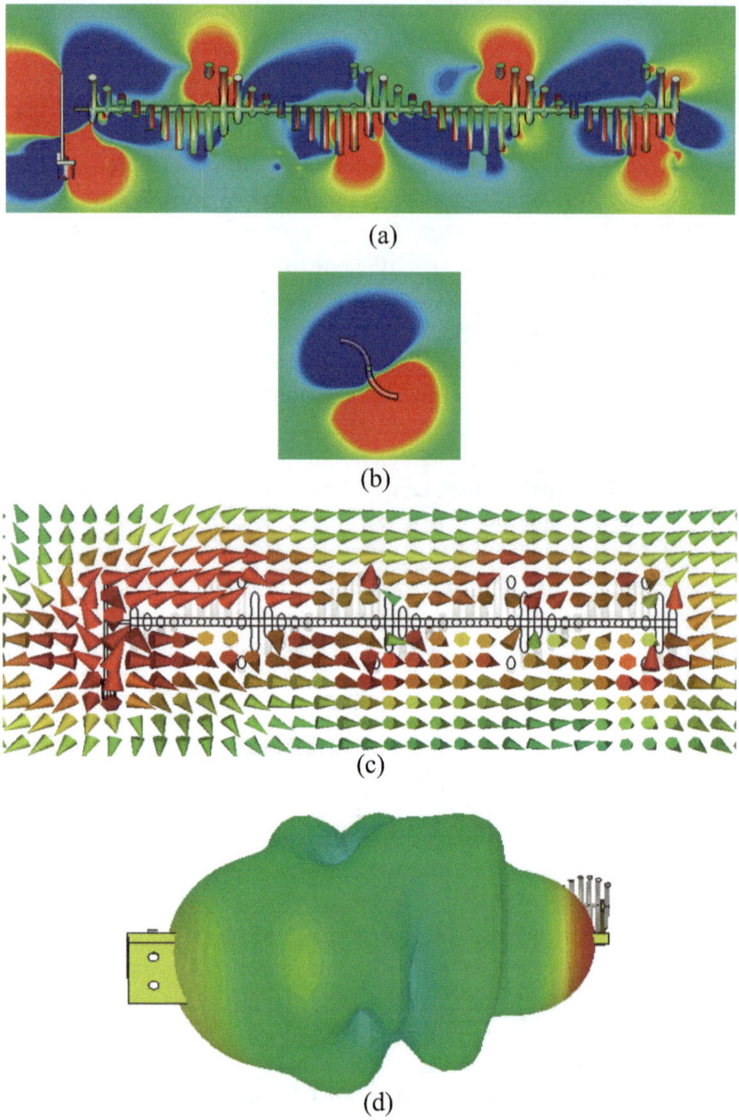

Fig. 6.15 Simulations result of end-fire antenna based on the rotated SSPPs structure (Str#3): **a**, **b** the electric field distribution in the section at the end of SSPPs structure and **c** the power flow; **d** the far-field at 2.45 GHz. Figure reproduced with permission from Ref. [3], © 2021 IEEE

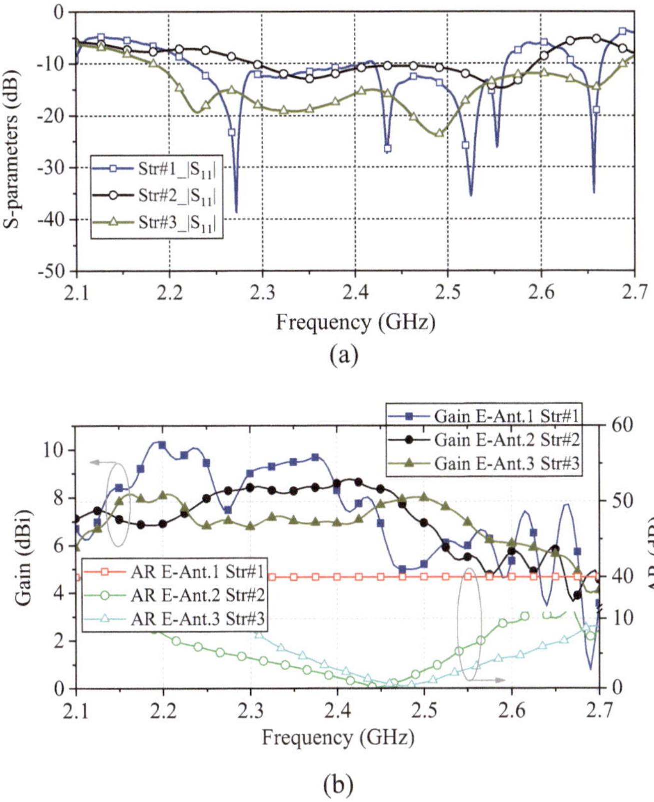

Fig. 6.16 Comparison between three kind end-fire antenna, E-Ant.1, E-Ant.2 ans E-Ant.3 based on Str#1, Str#2 and Str#3 respectively. **a** Simulated $|S_{11}|$ of end-fire antennas; **b** gain and axial ratios of end-fire antennas based these three stages of the SSPPs. Figure reproduced with permission from Ref. [3], © 2021 IEEE

6.2.2 Parametric Analysis to the End Fire Antenna Based on Str#3

To the end fire antenna based on the R-SSPPs structure Str#3 (E-Ant.3), the number of the s-shaped arms, rotated angle step $d\varphi$, and the bending radiators r_1 and r_2 of the s-shaped arm are the key parameters to the performances of antenna. The E-Ant.3 structure is complicated, and 3D print is used to process the antenna. The machining accuracy of 3D print is 2 mm, so we set the diameter of the s-shaped arm as $r_1 - r_2 = 3$ mm. And the key parameters are studied in the following.

(1) *N*

The curves of the gain and axial ratio of the end fire antenna E-Ant.3 varying with the number of elements N are shown in Fig. 6.17. It can be seen that the gain and

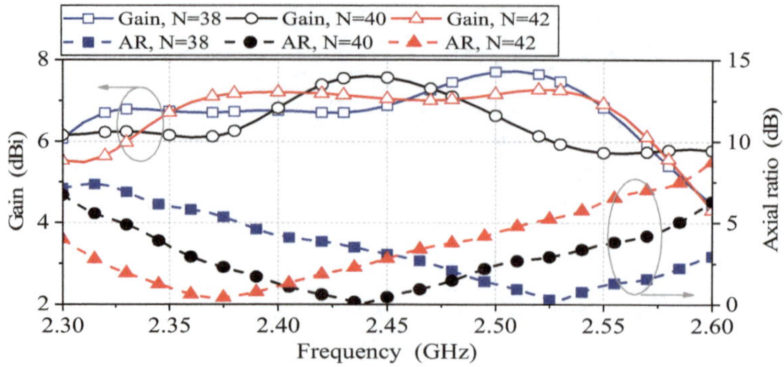

Fig. 6.17 Effects of N on the gains and axial ratio values of this proposed antenna. Figure reproduced with permission from Ref. [3], © 2021 IEEE

axial ratio of E-Ant.3 vary greatly with the number of elements N increasing from 38 to 42. When N = 38, the gain peak appears at 2.5 GHz. When N = 42, the gain peak appears near 2.4 GHz. When N = 40, the peak gain of the end fire antenna appears at 2.45 GHz, and the circular polarization (CP) band with AR < 3 dB is 2.36–2.53 GHz. It is obvious that when N increases, the peak gain and CP band of the antenna shift to lower frequency, but the peak gain and circularly polarized band of the antenna shift to higher frequency when N decreases.

(2) r_2

The influence of r_2 to the gain and axis ratio of E-Ant.3 is shown in Fig. 6.18. When r_2 = 13 mm, the antenna has stable high gain in a wide frequency band, and the corresponding circularly polarized bandwidth (AR < 3 dB) is 2.36–2.53 GHz. When r_2 deviates from 13 mm, the axial ratio of the antenna E-Ant. in

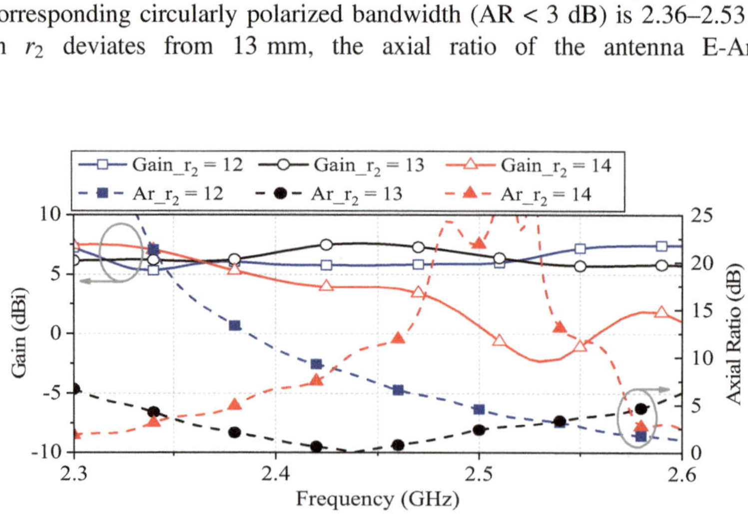

Fig. 6.18 Effects of r_2 on the gains and axial ratio values of this proposed end-fire antenna. Figure reproduced with permission from Ref. [3], © 2021 IEEE

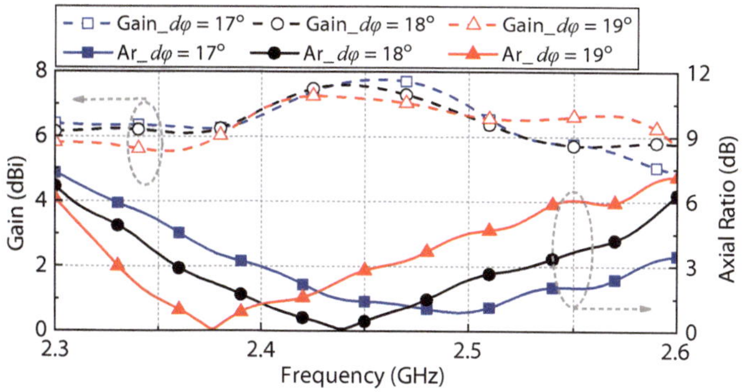

Fig. 6.19 Effects of $d\varphi$ on the gains and axial ratio values of this proposed end-fire antenna. Figure reproduced with permission from Ref. [3], © 2021 IEEE

2.36–2.53 GHz will deteriorate, and the frequency band with AR < 3 dB will rapidly shift to higher frequency or lower frequency.

(3) $d\varphi$

The influence curve of the rotation angle $d\varphi$ on the gain and axial ratio of end fire antenna E-Ant.3 is shown in Fig. 6.19. It can be seen that the change of rotation angle $d\varphi$ has little effect on the gain of E-Ant.3, but the circularly polarized frequency band (AR < 3 dB) will move to lower frequency with the increase of rotation angle $d\varphi$. According to the simulation results, we choose $d\varphi = 18°$.

After simulation and optimization, the optimized parameters of end fire antenna E-Ant.3 are shown in Table 6.1.

6.2.3 *Experiment Results Analysis of Endfire Antenna Based on Str#3*

On the basis of the optimal parameters obtained from the above simulation and optimization, we fabricated and tested the prototype of the end fire antenna based on R-SSPPs Str#3 according to the size data in Table 6.1. Figure 6.20 the fabricated prototype of the proposed end-fire antenna shows the prototype of the antenna. In addition, we also processed a resin bracket through 3D printing, which can facilitate the installation and testing of the antenna.

Table 6.1 The optimized antenna parameters of the end-fire antenna based on the odd mode of R-SSPPs (unit: mm)

Parameter	W_1	W_2	h	l_1	N	r_1	r_2
Value	6	2.6	34.5	208	40	16	13
Parameter	w						
Value	2						

Table reproduced with permission from Ref. [3], © 2021 IEEE

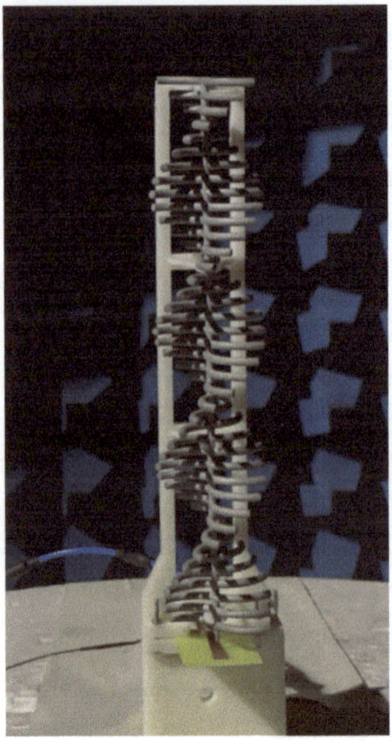

Fig. 6.20 The fabricated prototype of the proposed end-fire antenna. Figure reproduced with permission from Ref. [3], © 2021 IEEE

The return loss $|S_{11}|$ of the antenna prototype is measured by Agilent N8163A vector network analyzer, as shown in Fig. 6.21 Comparison of the simulated and measured $|S_{11}|$ of the proposed end-fire antenna based on the R-SSPPs in odd mode. It can be seen that the measured $|S_{11}|$ curve is basically consistent with the simulation result. The available impedance bandwidth of the antenna prototype is 2.185 ghz-2.68 GHz ($|S_{11}| < -10$ dB), and the relative bandwidth is 20.35%.

The measured patterns of E-Ant.3 antenna prototype at 2.4, 2.45 and 2.5 GHz are shown in Fig. 6.22. It can be seen that the antenna works in the end fire mode at

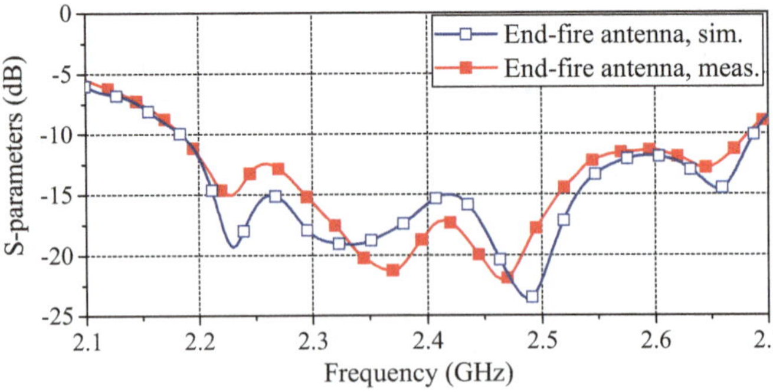

Fig. 6.21 Comparison of the simulated and measured $|S_{11}|$ of the proposed end-fire antenna based on the R-SSPPs in odd mode. Figure reproduced with permission from Ref. [3], © 2021 IEEE

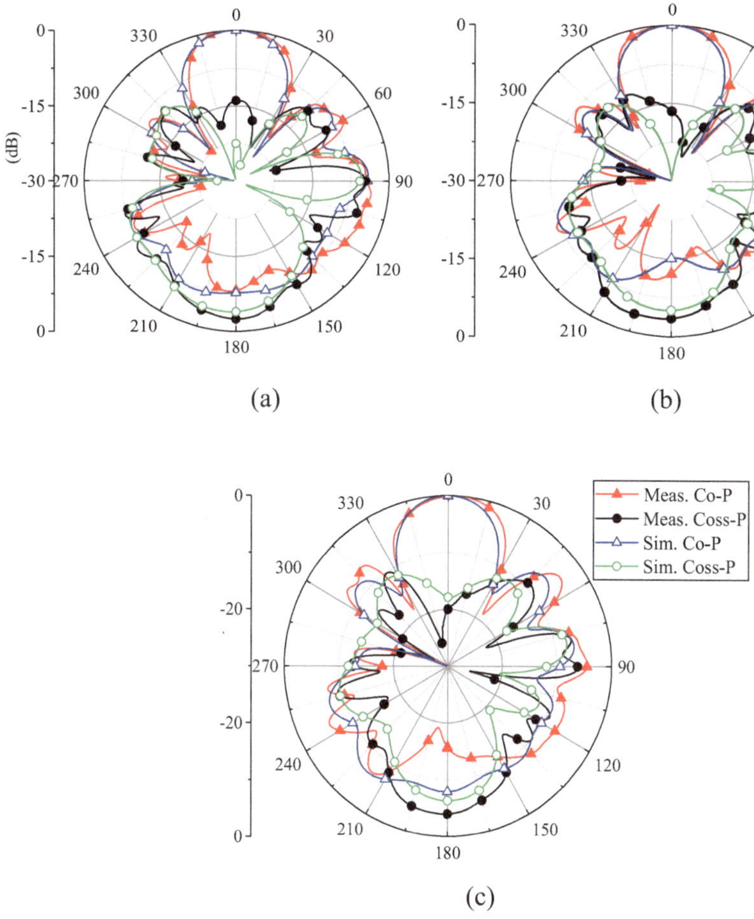

Fig. 6.22 Simulated and measured radiation patterns of the proposed end-fire antenna based on the R-SSPPs in odd mode, **a** 2.4 GHz; **b** 2.45 GHz; **c** 2.5 GHz. Figure reproduced with permission from Ref. [3], © 2021 IEEE

these three frequencies, and the measured patterns of co-polarization and cross polarization are consistent with the simulation results respectively. The measured and simulated main beam of the main polarization pattern points to 0° and it is more than 10 dB higher than the cross polarization level.

The measured and simulated gain and axial ratio curves of E-Ant.3 antenna prototype are shown in Fig. 6.23. The frequency band with measured AR < 3 dB is 2.435−2.525 GHz, and the measured gain of antenna in this frequency band is 6–7.43 dBi, which is relatively stable. Compared with the simulation results, the measured circularly polarized frequency band with axial ratio less than 3 dB is shifted to highe frequency, which is mainly caused by the errors in antenna processing and rinstallation.

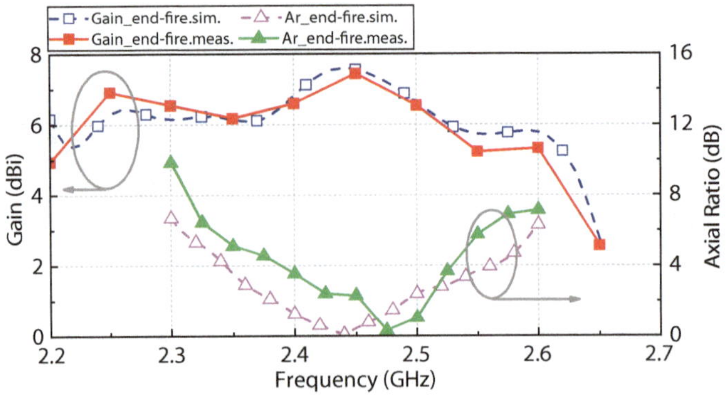

Fig. 6.23 Comparison of the simulated and measured gains and AR of the proposed end-fire antenna based on the R-SSPPs in odd mode. Figure reproduced with permission from Ref. [3], © 2021 IEEE

6.3 Omnidirectional Broadside Antenna Based on Rotated Spoof Surface Plasmon Polaritons

In Sect. 6.1, we have shown the transmission and radiation characteristics of the R-SSPPs structure (Str#3) in even modes when it is radiated by the plane wave laterally. In this section, an omni-directional radiation antenna based on R-SSPPs structure (Str#3) is realized by introducing an actual feed structure to replace the plane wave in Fig. 6.10. The specific antenna model and performance will be discussed in detail below.

6.3.1 Antenna Model and Simulation Verification

The front view of the broadside radiation antenna based on the R-SSPPs (Str#3) and the details of its feed structure are shown in Fig. 6.24, which is called B-Ant.3. SMA connector is shown in gray part in the figure, printed metal is blue part, and medium is orange part. The broadside radiation antenna mainly includes R-SSPPs transmission and radiation structure and CPW feed structure. The R-SSPPs Str#3 and CPW are connected by a part of metal transition structure. The gradual change floor of CPW is located on both sides of trapezoidal metal transmission line. CPW and metal floor are printed on one side of FR-4 dielectric plate, and the thickness of FR-4 dielectric plate is 1 mm. The dielectric constant of the FR-4 plate is 4.4 and the loss tangent is 0.025. The specific dimension parameters are shown in Fig. 6.24. The main parameters in Str#3 of R-SSPPs structure are r_1 = 16 mm, r_2 = 13 mm.

When CPW is introduced to feed the R-SSPPs structure instead of the side incident plane wave, the electromagnetic signal first enters the CPW through SMA

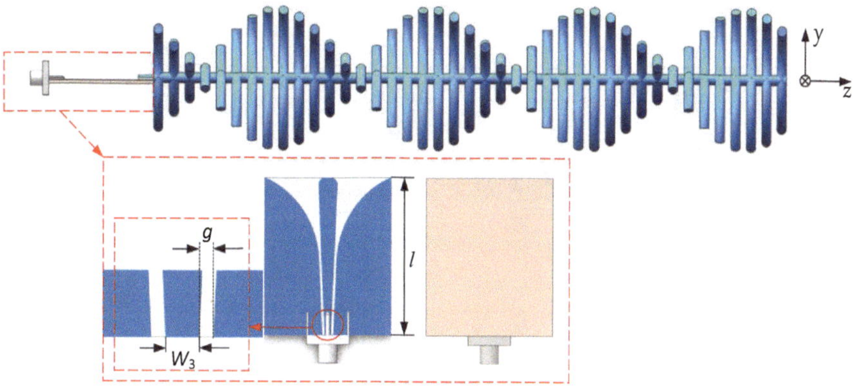

Fig. 6.24 The front view of the broadside antenna based on Str#3 and the details of the feed. Figure reproduced with permission from Ref. [3], © 2021 IEEE

connector, and then enters the R-SSPPs structure along the strip line/metal arm. The current distribution on the R-SSPPs structure fed by CPW is similar to that of R-SSPPs in quasi even mode in Sect. 6.1, and the current direction of the upper and lower parts of the s-shaped arm of R-SSPPs is opposite. The electric field distribution of the broadside radiation antenna at 2.45 GHz is shown in Fig. 6.25a, b. It is obvious that the electric field of the antenna is basically symmetrical about the Z-axis in the *xoz* plane in Fig. 6.25a and approximately symmetrical about the s-shaped arm in the *xoy* sectionin Fig. 6.25b, which indicates that the electromagnetic wave entering the R-SSPPs structure through CPW excites quasi even modes in the R-SSPPs structure. The simulation results of the power flow distribution of B-Ant.3 antenna is shown in Fig. 6.25c. It is obvious that after the electromagnetic energy enters the R-SSPPs structure Str#3 from CPW, it propagates to the right along the *z* axis of R-SSPPs and radiates continuously from the s-shaped arm to the free space in the *xoy* plane, forming the quasi omnidirectional radiation pattern in the *xoy* plane, as shown in Fig. 6.25d. Obviously, these simulation results of the broadside rdiation antenna B-Ant.3 based on the R-SSPPs even mode are consistent with the even mode characteristics of the R-SSPPs structure discussed and analyzed in Sect. 6.1.

For the sake of comparison, Str#1 and Str#2 are also excited and fed from CPW respectively, which are called B-Ant.1 and B-Ant.2. And the arm lengths of Str#1 and Str#2 are $l_e = 45.5$ mm. The simulation results of return loss of the broadside radiation antenna based on these three kinds of SSPPs structure are shown in Fig. 6.26. Obviously, the impedance bandwidth ($|S_{11}| < -10$ dB) of the even/quasi even mode broadside radiation antenna based on R-SSPPs Str#3 (E-Ant.3) can cover 2.352–2.7 GHz, which is larger than that of.

The avialiable impedance of the broadside radiation antenna B-Ant.1 covers 2.356–2.627 GHz and B-Ant.2 covers 2.183–2.7 GHz. Figure 6.26 shows the simulated gain curves of these three broadside radiation SSPPs antennas too, and

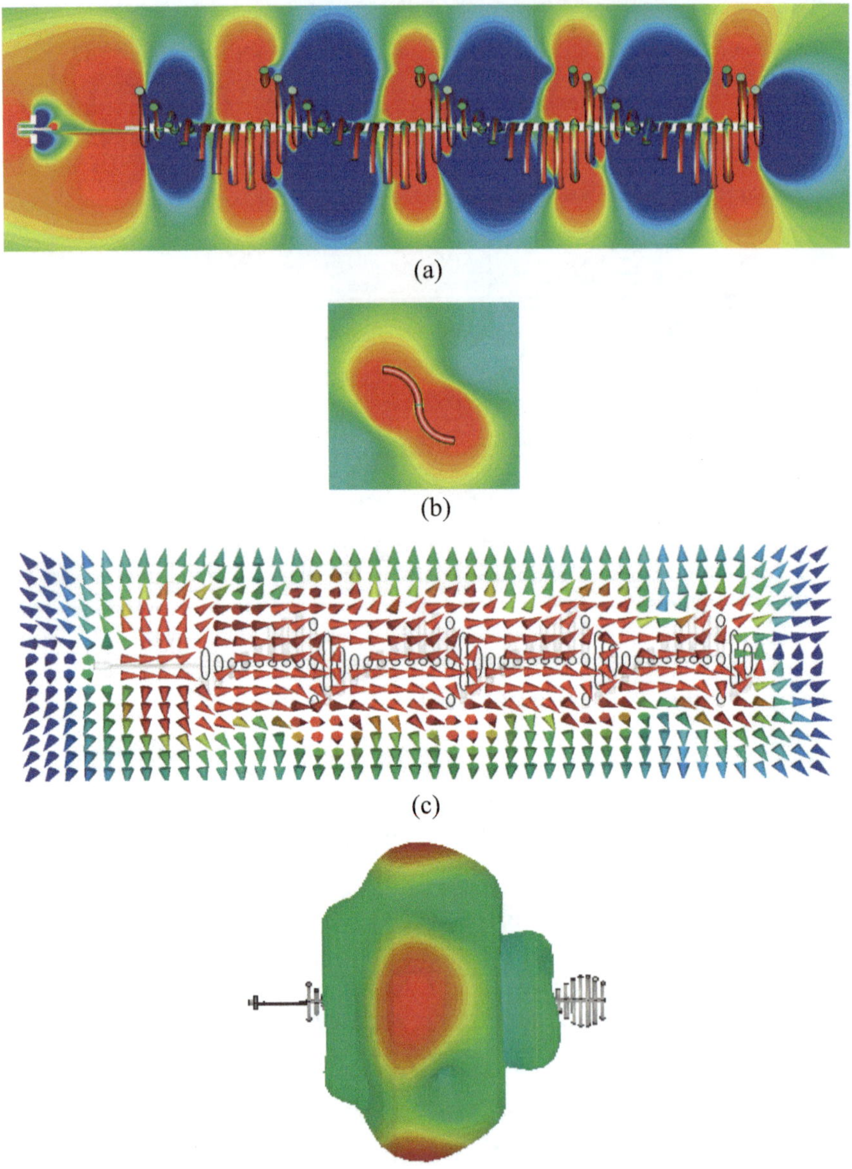

(a)

(b)

(c)

(d)

Fig. 6.25 Simulation results of the broadside antenna, B-Ant.3, based on the rotated SSPPs structure (Str#3): **a**, **b** the electric field distribution in the section at the end of SSPPs structure and **c** the power flow; **d** the simulation far-field patterns at 2. Figure reproduced with permission from Ref. [3], © 2021 IEEE

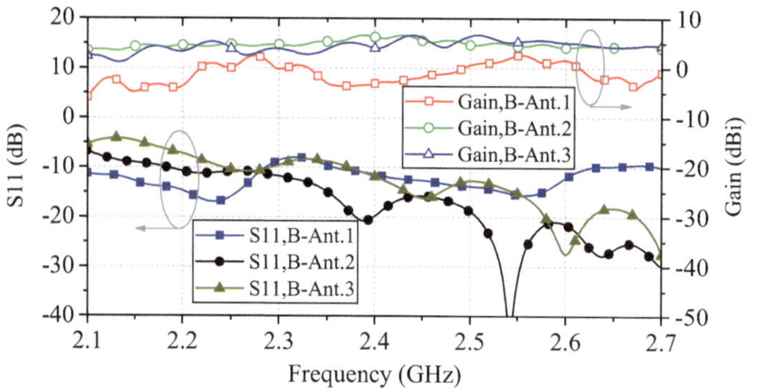

Fig. 6.26 Comparison $|S_{11}|$ and gain between three kind broadside antenna B-Ant.1, B-Ant.2 and B-Ant.3 based on Str#1, Str#2 and Str#3., and they all are fed similar as Fig. 6.24. Figure reproduced with permission from Ref. [3], © 2021 IEEE

the gain is −2.54–2.637 dBi to B-Ant.1, 4.17–6.645 dBi to B-Ant.2, and 3.81–6.77 dBi to B-Ant.3, respectively. It is noted that the B-Ant.2 and B-Ant.3 based on R-SSPPs structure perform higher gain than the B-Ant.1 based on tradiational dual arm SSPPs structure Str#1.

6.3.2 Parameter Analysis

In the previous simulation process, it is found that the performance of broadside radiation antenna B-Ant.3 based on R-SSPPs structure (Str#3) is also affected by several key parameters, such as the number of s-shaped elements N, the rotation angle $d\varphi$ and the inner and outer radius s-shaped arm r_1 and r_2. Similar to endfire antenna, considering the machining accuracy of 3D printing, in this section, we set the diameter of s-shaped arm as $r_1 - r_2 = 3$ mm. The influence of these parameters are analyzed in detail.

The influence of the change of the number of elements N on the pattern of the broadside radiation antenna based on R-SSPPs quasi even mode is shown in Fig. 6.27. It can be seen that when the number of elements $N = 38$, the main beam of the antenna pattern points to near $\theta = 110°$ and has obvious side lobes; when the number of elements $N = 40$, the main beam of the antenna pattern is relatively clean and points to near $\theta = 90°$; when the number of elements $N = 42$, the main beam of the antenna pattern points to near $\theta = 90°$ and has obvious side lobes too. Clearly, when the number of elements N deviates from 40, the main beam of the antenna broadside radiation pattern will deviate from the direction of $\theta = 90°$ or near sidelobes will appear, and the gain of the main beam will also decrease significantly. So, we choose $N = 40$.

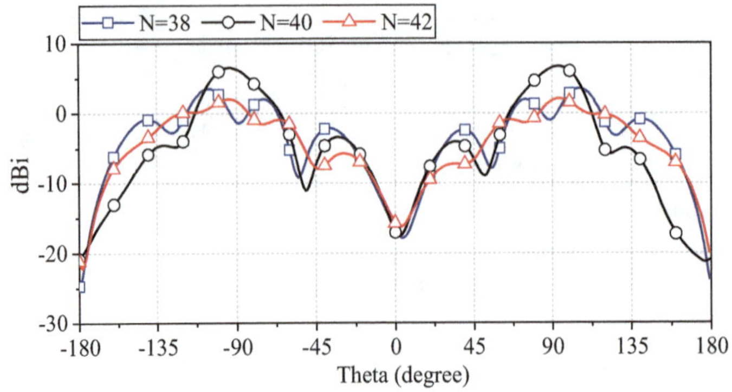

Fig. 6.27 Effects of N on the radiation pattern of this proposed antenna. Figure reproduced with permission from Ref. [3], © 2021 IEEE

In addition, r_2 = 12 mm, 13 mm and 14 mm, the two-dimensional patterns of the broadside radiation antenna B-Ant.3 based on R-SSPPs structure in quasi even mode in the *yoz* plane is shown in Fig. 6.28. As you can see, r_2 has the same effect on the antenna pattern as the number of s-shaped elements *N*. When r_2 is not equal to 12 mm, the main beam of antenna sidelobe pattern will deviate from the direction of $\theta = 90°$ and the gain of main beam will decrease.

The influence of the change of another parameter $d\varphi$ on the radiation pattern of the broadside radiation antenna B-Ant.3 is shown in Fig. 6.29. It can be seen that the change of $d\varphi$ has some small influence on the radiation pattern of the antenna, but has less influence on the direction of the main beam, which can be almost ignored. However, when $d\varphi$ deviates from 18°, the gain of antenna will decrease obviously no matter $d\varphi$ increasing or decreasing.

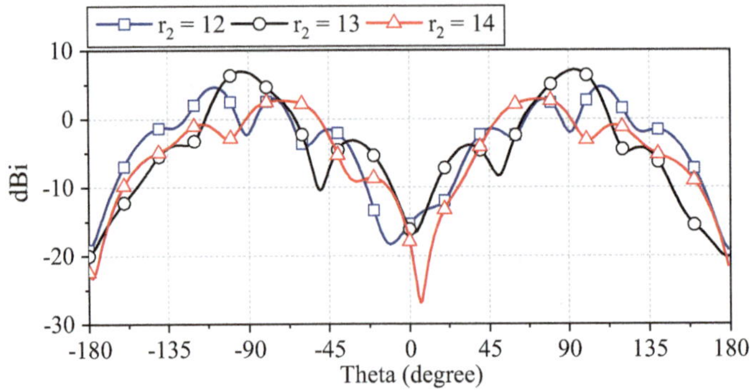

Fig. 6.28 Effects of r_2 on the radiation pattern of this proposed antenna. Figure reproduced with permission from Ref. [3], © 2021 IEEE

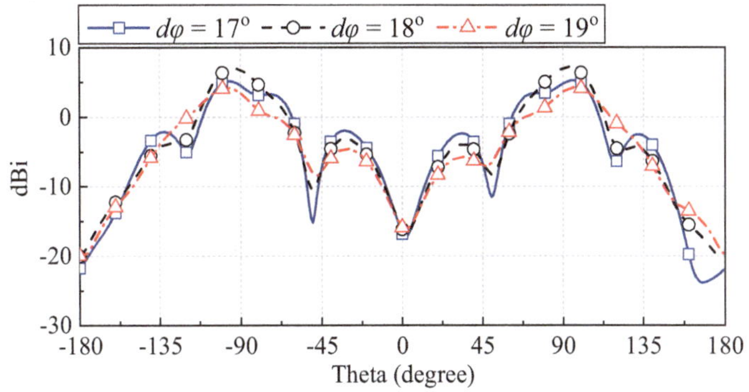

Fig. 6.29 Effects of $d\varphi$ on the radiation pattern of this proposed antenna. Figure reproduced with permission from Ref. [3], © 2021 IEEE

After the simulation and optimization of the main parameters of the broadside radiation R-SSPPs antenna, the optimal parameters of the broadside radiation antenna B-Ant.3 are shown in Table 6.2.

6.3.3 Experiment Results Analysis to Broadside Antenna Based on Str#3

According to the optimal parameters in Table 6.2, we have practically processed the final optimized broadside radiation antenna prototype and measured in chamber. Figure 6.30 shows the real broadside radiation antenna. Similarly, 3D printed resin bracket support is used for installation and test.

The measured reflection coefficient $|S_{11}|$ of the broadside radiation antenna is shown in Fig. 6.31. It can be seen that the measured results are in good agreement with the simulation result. The measured impedance bandwidth of the broadside radiation antenna ($|S_{11}| < -10$ dB) from 2.35 to 3 GHz, the relative impedance bandwidth is about 23.2%, which has wide band effect.

The far-field patterns of the antenna in the measured *xoy* plane at 2.4, 2.45 and 2.5 GHz are shown in Fig. 6.32. Obviously, in the *xoy* plane, the pattern is quasi omnidirectional, and means that the antenna works in the broadside radaition mode. Figure 6.33 shows the measured gain curve of the broadside radaition antenna. It

Table 6.2 The optimized antenna parameters of the broadside antenna based on the even mode of R-SSPPs (unit: mm)

Parameters	W_3	g	l	l_1	N	r_1	r_2
Value	1.4	0.6	34	208	40	16	13
Parameters	w						
Value	2						

Table reproduced with permission from Ref. [3], © 2021 IEEE

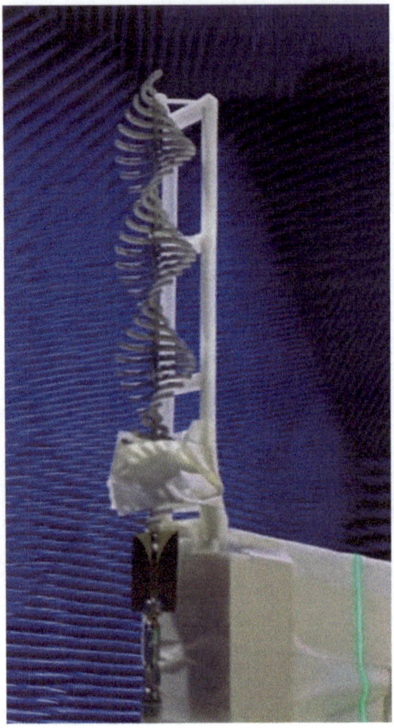

Fig. 6.30 The fabricated prototype of the proposed broadside antenna. Figure reproduced with permission from Ref. [3], © 2021 IEEE

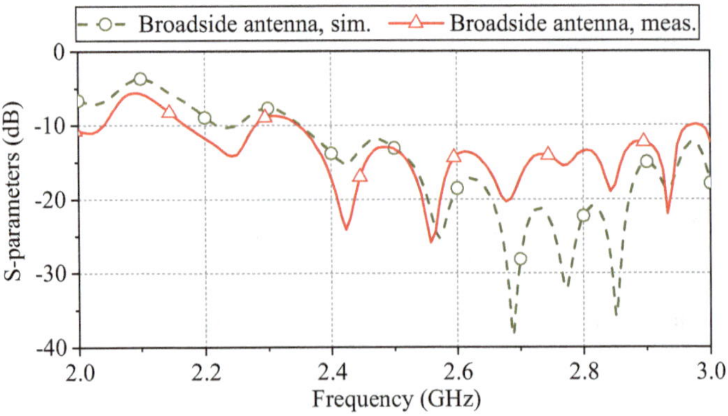

Fig. 6.31 Comparison of the simulated and measured $|S_{11}|$ of the proposed broadside antenna based on the R-SSPPs in even mode. Figure reproduced with permission from Ref. [3], © 2021 IEEE

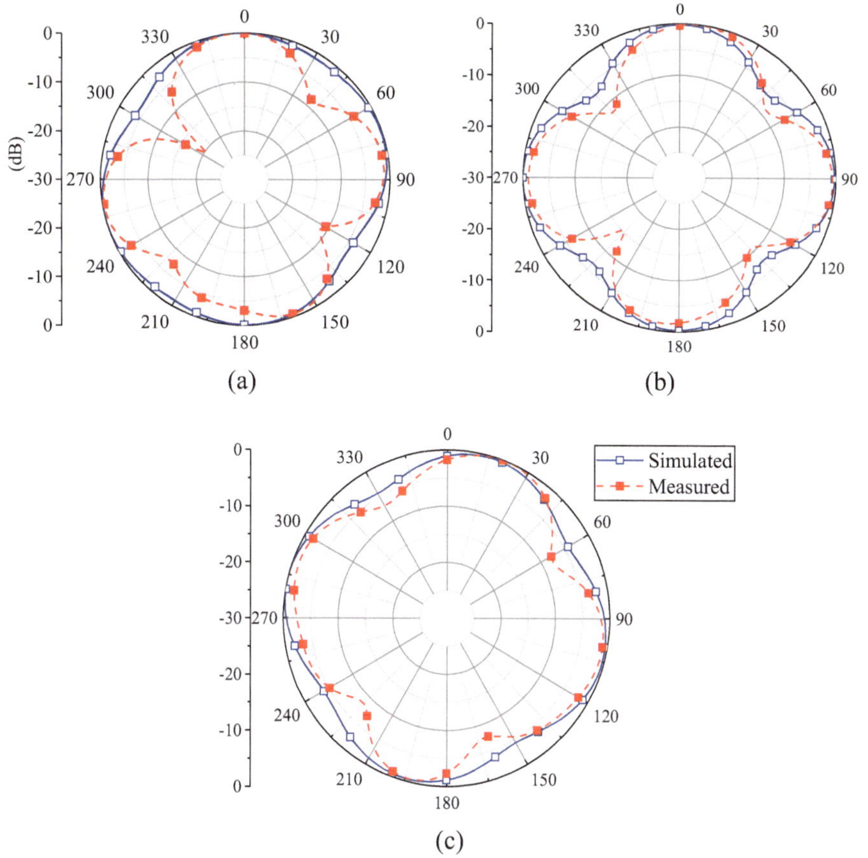

Fig. 6.32 Simulated and measured radiation patterns of the proposed broadside antenna based on the R-SSPPs in even mode. Figure reproduced with permission from Ref. [3], © 2021 IEEE

can be seen that measured gain within the impedance bandwidth ($|S_{11}| < -10$ dB) of the antenna is 2–6 dBi, which is basically consistent with the simulation trend.

On the one hand, the material of 3D printed R-SSPPs structure is aluminum alloy, and its conductivity is worse than that of copper or silver. The welding of aluminum and copper microstrip lines is also difficult. In addition, the machining error is another reason to lead to the difference between the actual antenna gain and the simulation.

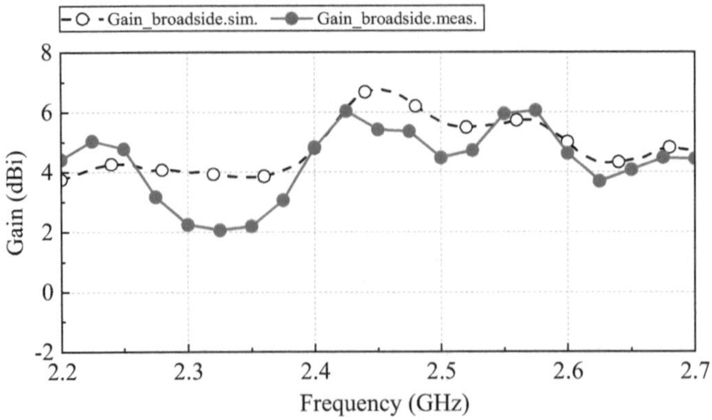

Fig. 6.33 Comparison of the simulated and measured gains of the proposed broadside antenna based on the R-SSPPs in even mode. Figure reproduced with permission from Ref. [3], © 2021 IEEE

6.4 Result Comparison

These two rotated SSPPs antennas (E-Ant.3 and B-Ant.3) are compared with other similar antennas, such as the helical antennas and other SSPPs antennas in Fig. 6.33. Both the rotated SSPPs antennas (E-Ant.3 and B-Ant.3) show novel radiation performances. The E-Ant.3 shows a stable end-fire beam with good CP character and high gain in the odd mode. The B-Ant.3 performs the quasi-omnidirectional radiation in the quasi-even mode over a wide frequency band. It means that the rotated SSPPs antennas have a good promotion to the wireless transmission (Table 6.3).

6.5 Summary of This Chapter

The rotated SSPPs structure is presented, which can transmit odd mode or quasi-even mode electromagnetic wave. The plane wave can excite the odd mode running wave in the rotated SSPPs structure to end-fire radiation with CP characteristics, or quasi even mode running wave to omnidirectional broadside radiation respectively. The end-fire CP antenna based on the R-SSPPs structure is designed, and the proyotype is manufactured, which is fed by monopole. The measured impendence band with $|S_{11}| < -10$ dB is 2.185–2.68 GHz, and the measured CP range is 2.41–2.52 GHz with gain 6–7.43 dBi. The broadside radiation antenna based on the R-SSPPs structure is designed, and the proyotype is manufactured too, which is fed by SMA connector and CPW along z axis to R-SSPPs structure.

Table 6.3 Comparisons between the proposed antennas and other works

Ref.	Basic model	Mode	Frequency (GHz)	Pattern	Polarization	Gai (dBi)	Angular width (3 dB)
[4]	Spiral	Axial mode	/	End-fire	CP	/	
[4]	Spiral	Normal mode	/	Broadside	LP	/	
[5]	SSPPs	Odd mode	6.5–11.5	Endfire	LP	7.5–9.2	
[6]	SSPPs	Even mode	0.302–0.335	Omnidirectional	LP	0.9–1.76	28.8°
This work	Rotated SSPPs	Odd mode	2.185–2.68	Endfire	CP@ (2.41–2.52 GHz)	6–7.43	33°–43°
		Even mode	2.35–3	Quasiomnidirectional	LP	2–6	22°–44°

Table reproduced with permission from Ref. [3], © 2021 IEEE

The measured pattern is quasi-omnidirectional in side direction, and the measured impedance band of the broadside radiation antenna prototype is in 2.35–3 GHz (in quasi-even mode) with gain 2–6 dBi.

The proposed R-SSPPs structure guarantees the good performance of circular polarization of the antenna in odd mode and broadside radiation in quasi-even mode. Different modes, polarizations and radiation patterns in the R-SSPPs structure may bring a good promotion of multi-functions in future wireless communication system.

References

1. K. Zhuang et al., Pattern reconfigurable antenna applying spoof surface plasmon polaritons for wide angle beam steering. IEEE Access **7**, 15444–15451 (2019)
2. K. Wang et al., in *Pattern Reconfigurable Antenna Applying Spoof Surface Plasmon Polaritons*. 2019 13th European Conference on Antennas and Propagation (EuCAP), Krakow, Poland, 2019, pp. 1–2
3. C. Xie et al., in *Study on the Rotated SSPPs Structure and its Applications in Antenna*. IEEE Transactions on Antennas and Propagation
4. J.D. Kraus, R.M. Marhefka, *Antennas: For All Applications*, 3rd edn. (McGraw-Hill, New York, 2008), pp. 222–303
5. A. Kandwal, Q. Zhang, X.L. Tang, L.W. Liu, G. Zhang, Low-profile spoof surface plasmon polaritons traveling-wave antenna for near-endfire radiation. IEEE Antenn. Wirel. Pr. **17**(2), 184–187 (2018)
6. D. Wang, J. Geng, K. Wang, H. Zhou, J. Li, X. Liang, W. Zhu, R. Jin, A high-efficiency broadband omnidirectional UHF patch antenna applying surface plasmon polaritons for handheld terminals. IEEE Antenn. Wirel. Pr. **17**(2), 283–286 (2018)
7. Zhuang et al., Spoof surface plasmon polaritons pattern reconfigurable antenna for wide-angle coverage. Proc IEEE Int Symp Antennas Propag USNC/URSI Nat Radio Sci Meeting Jul. 2018, pp. 983–984

Chapter 7
Irregular Shaped Spoof Surface Plasmon Polaritons Antenna

Junping Geng, Chaofan Ren, and Xuxu Cheng

7.1 Introduction

Compact antenna with low profile is urgently needed in modern wireless communication system. Especially in moving vehicle communication or unmanned aerial vehicles (UAVs) communication, it doesn't only need the directional antenna, tangent directional antenna and omnidirectional antenna, but also combine the low-profile conformal antenna into the shell to high load bearing.

The surface wave technology has been used to design the low-profile antenna. In [1, 2], the meta-surface is used to design low-profile antenna. [3] introduced the metasurface of rectangular metallic patch array to excite the TE surface wave. It achieved directional pattern with peak gain of 10.3 dBi and a low profile of 0.06 λ_0. In [4], the AMC surface is designed to realize the low profile wideband linearly polarized antenna. In [5, 6] the HIS surface is inserted to realize the low-profile too. All above low-profile antennas with surface wave technology are directional radiation with parallel polarization, and they are difficult to radiate omnidirectional or tangential wave with vertical polarization. In [7], a tapered grounded ceramic slab is proposed to transform the guided surface wave into radiated space wave over a wide frequency range. But the above compact antennas in the low frequency band, such as UHF band, are still too large concerning conformal applications with surface wave technology.

Spoof surface plasmon polaritons (SSPPs) antenna is another kind of surface wave structure, which is a good choice for compact antenna design due to its subwavelength characteristics, high gain in end-fire mode and omnidirectional radiation.

J. Geng (✉) · C. Ren · X. Cheng
Department of Electronic Engineering, Shanghai Jiao Tong University, Shanghai 200240, China
e-mail: gengjunp@sjtu.edu.cn

J. Geng et al., *Spoof Surface Plasmon Polaritons Antenna*, Modern Antenna,
https://doi.org/10.1007/978-981-16-4721-5_7

Spoof surface plasmon polaritons [8] structure has aroused tremendous interest among modern researchers since it could operate with sub-wavelength size which is to compact designing. It was firstly considered as an outstanding form of transmission line, for the advantages of low loss and high efficiency [9, 10]. Further study in [11, 12] shows that SSPPs structure propagates SPPs wave in odd and even modes. Later, SSPPs structure was extended into antenna by introducing gradient structure [13] or inserting radiation elements [14], such as dielectric resonators [15].

SSPPs antennas are basically focused on end-fire antennas and broadside antennas, both of which are normally planar structure printed on dielectric. Kandwal et al. [16] proposes a low profile odd-mode SSPPs antenna with co-planar waveguide (CPW), which is a typical configuration of end-fire SSPPs antenna. In [17], the difference of end-fire radiation beam according to different equivalent aperture of single-sided SSPPs and double-sided SSPPs is analyzed. The SSPPs antenna generates end fire radiation beam when it operates in odd mode and vice versa. Wu and Shen [18] design a broadside leaky wave antenna by introducing hollowed out SSPPs block. A dual port even mode SSPPs antenna is presented in [19]. It radiates broadside beam when two ports are fed at the same time. Zhuang [20] utilizes PIN diodes to switch between odd and even mode realizing pattern reconfiguration between end-fire and broadside radiation.

Besides, there are many various SSPPs antennas that bring new insight and differences into SSPPs antenna design. Han and Li [21] present a multibeam antenna based on SSPPs mode coupling and focus on realizing high gain omni-directional antenna by using several SSPPs coupling units to form circular array. Fan [22] analyzes the principle of frequency scanning radiation SSPPs antenna and design an antenna correspondingly with gradient metasurface operating in 8.8–10.7 GHz with maximum gain of 13.1 dB. Kong [23, 24] proposed a leaky wave SSPPs antenna with the groove depth of SSPPs unit cell modulated and therefore it generates omnidirectional broadside radiation beam with 12 dB gain and frequency-controlled beam scanning angle from 67.7° to 35.6°.

Even though the odd mode SSPPs structure was mounted on the ground to realize end fire radiation with SSPP arms being vertical to the ground, its profile at lower frequency band is still relatively large for conformal design, especially to combine the low-profile conformal antenna into the shell to load bearing in high speed vehicle.

7.2 Low Profile Vertically Polarized Endfire Antenna Based on Spoof Surface Plasmon Polaritons

A SSPPs terminal antenna fed by asymmetric gap coupling, is propoased in [25], as shown in Fig. 7.1. It is a typical end-fire spoof surface plasmon polaritons antenna, but the profile is highe when it works in vertical polarization. Here we introduce the ground into the SSPPs structure, and cut the SSPPs structure into two parts. And the mirror of the ground just takes the place of half part of the SSPPs structure. [26]

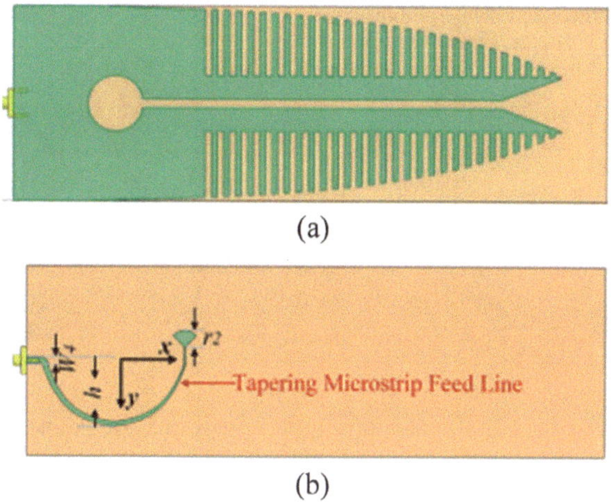

Fig. 7.1 The end-fire antenna based **a** front view, **b** back view

The geometry structure of the proposed antenna is given in Fig. 7.2. It includes the feeding microstrip line, the SSPPs transmission structure in the front and the radiation part. The odd SSPPs mode is stimulated and transferred along the SSPPs structure, and leaked into free space from the end of the SSPPs structure. The dielectric substrate of the antenna is FR-4 with relative dielectric constant is 4.3 and thickness is 1 mm. The electric height of antenna is 0.18λ @1.5 GHz, which is quite low profile for a vertically polarized antenna.

The simulated return loss curve is shown in Fig. 7.3. It is clear that the aviliable impedance bandwidth is from 1.408 to 1.748 GHz and from 1.794 to 1.922 GHz with return loss being less than −10 dB.

The simulated far fields @1.5, 1.7 and 1.95 GHz are given in Fig. 7.4. They all showed wide end fire beam perfoamances within the whole working bandwidth.

In Fig. 7.4, the realized gain reaches up to 10 dBi and no less than 6 dBi. And the 3 dB beamwidth is up to 250 degrees.

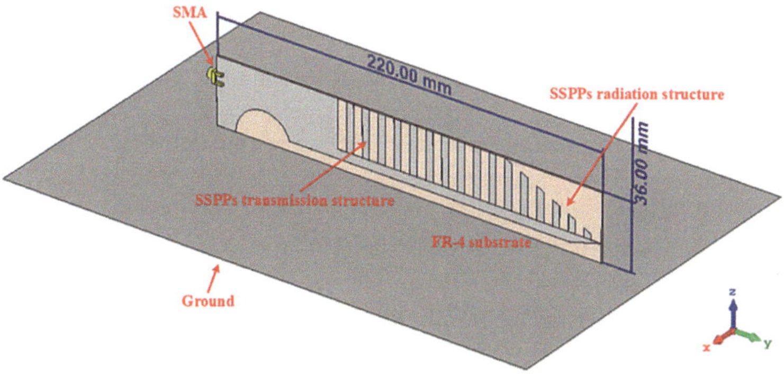

Fig. 7.2 Geometry structure of antenna. Figure reproduced with permission from Ref. [26], © 2020 IEEE

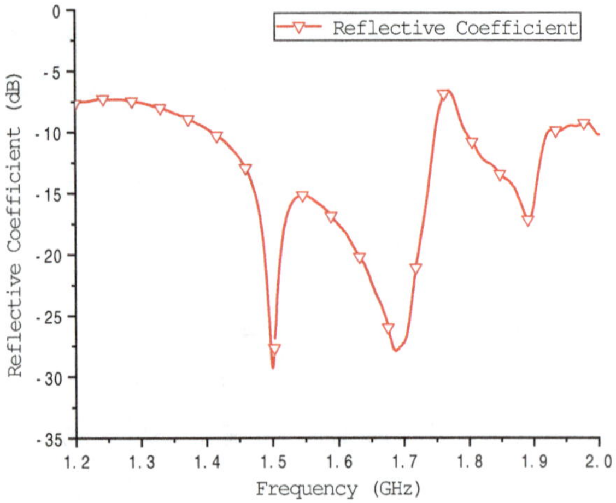

Fig. 7.3 Simulated reflective coefficient. Figure reproduced with permission from Ref. [26], © 2020 IEEE

7.3 Low Profile M-shaped Spoof Surface Plasmon Polaritons Terminal Antenna and Array

In this section, a method to decrease the profile of the vertical polarized antenna with m-shaped SSPPs structure is proposed. Firstly, the ground is inserted in the middle of the tradional double side SSPPs structure, and the ground is vertical to the SSPPs plane, and generates the anti-symetrical mirror image of the upward single side SSPPs structure, so that the upward single side SSPPs structure and its mirror image construct the odd mode dual side SSPPs structure. So, the profile of the proposed SSPPs structure is cut into half. Secondly, the end of the single arm of the SSPPs structure is split into two branches to be T-shaped arm, and the vertical part of the arm of the SSPPs structure with lower profile to provide odd mode dominated vertical polarization. The top horizontal part of the arms are symetrical to be able to support even mode to widen the beam of the end-fire radiation. Thirdly, the T-shaped arms are further bended into m-shaped structure to realize much lower profile and small size while keeps dominant odd mode performances. The mainly vertical polarized SSPPs end-fire antennas with m-shaped arms can work around 480 MHz and 1.5 GHz. The proposed antenna are mainly constructed with metal bars by mechanical fabrication, besides good radiation performances, it is fit for implanting in the shell of vehicle. In the following, the dispersive characteristics of three kind SSPPs structures, traditional SSPPs, T-shaped SSPPs unit and m-shaped SSPPs unit are discussed. Then the antenna principle, designing method and simulation results are discussed in details. The prototype of the optimized SSPPs antenna based on the m-shaped SSPPs structure is manufactured. The experiment results are in consistent with the simulations.

Fig. 7.4 Radiation pattern at different frequency. Figure reproduced with permission from Ref. [26], © 2020 IEEE

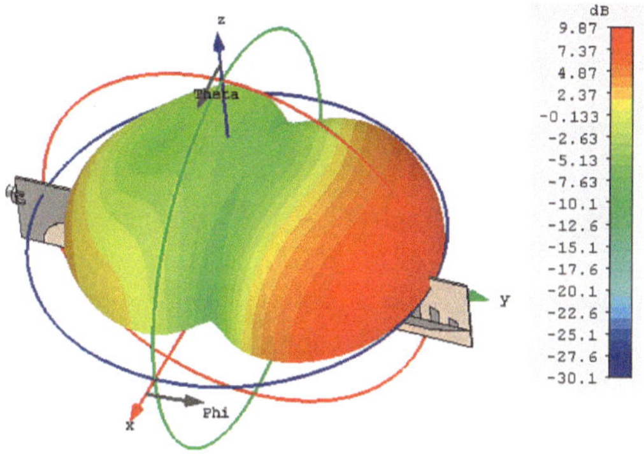

(a)1.5GHz

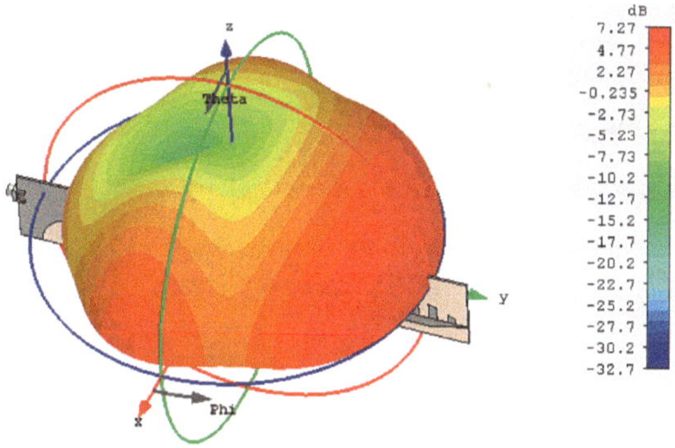

(b)1.7GHz

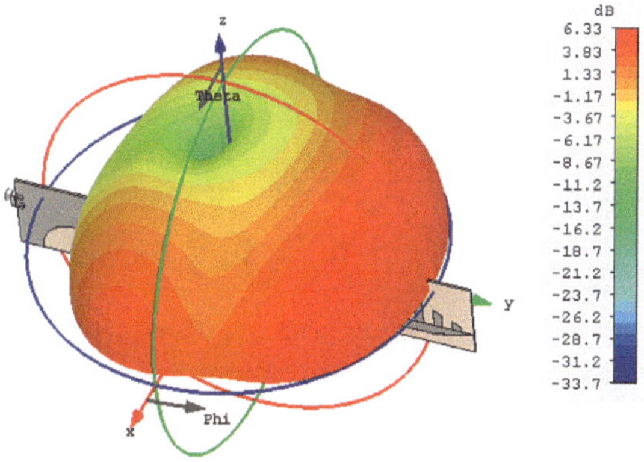

(c)1.96GHz

7.3.1 M-shaped Spoof Surface Plasmon Polaritons Structure Analysis

A. *M-shaped SSPPs Structure*

In [25], a good end-fire the SSPPs antenna with a center slot is investigated, consider half part of the SSPPs structure which like a comb, and set the single side SSPPs structure with a slot s_1 beneath the structure as the basic SSPPs unit 1 in Fig. 7.5a. It can support good odd mode SSPPs wave transmission performance too. To decrease the profile of the SSPPs structure and keep work band still in lower band of hundreds of MHz, the end of the single arm of the SSPPs structure is split into two branches to be T-shaped arm, which is set as unit 2 in Fig. 7.5b. The height h_2 decreases, but the bended end arm length w_2 is increased to maintain available electric length for designated frequency band. To further decrease the profile to compact the SSPPs structure, the arms of the SSPPs structure are bended twice to be the m-shape, which is clled SSPPs structure unit 3 in Fig. 7.5c. Its profile height h_3 and the width w_3 can be reduced. The folded end wings with length e are added to compensate the whole electrical length.

B. *Dispersive and Transmission Characters of SSPPs*

Dispersion character is the key character to SSPPs structure. The dispersion curve of the traditional SSPPs comb structure could be calculated by the equation [27, 28]:

$$\beta = k_0\sqrt{1 + \left(\frac{a_1}{p_1}\right)^2 \tan^2(k_0 h_1)} \tag{7.1}$$

Here, the wave number in free space is $k_0 = c/\omega$, and c is light speed, h_1 is the height of the SSPPs arm, a_1 is the width of the slot of SSPPs unit and p_1 is the periodicity of the SSPPs unit. It is noted that the cut off frequency of the SSPPs structure is negatively related to h_1 According to Eq. (7.1).

In aother hand, the total number of SSPP modes propagating in the SSPPs structure could be calculated by the following equation [27]:

$$n = 1 + int(d_1 \lceil p_1) \tag{7.2}$$

Here, the symbol '$\lceil$' refers to round down.

To work in UHF band of hundreds of MHz, the original parameters of three kinds of SSPPs structure unit 1, unit 2 and unite 3 are given in Table 7.1. And the dispersive curves are shown in Fig. 7.5d.

To unit 1, the cut-off frequency of the 1st SSPP mode is close to 500 MHz, and the cut-off frequency of the 2nd SSPP mode is close to 2 GHz. There could support up to 7 SSPP modes in this case. To unit 2, the cut-off frequencies in the 1st mode and the 2nd mode are 540 MHz and 2.03 GHz respectively. To unit 3, the cut-off

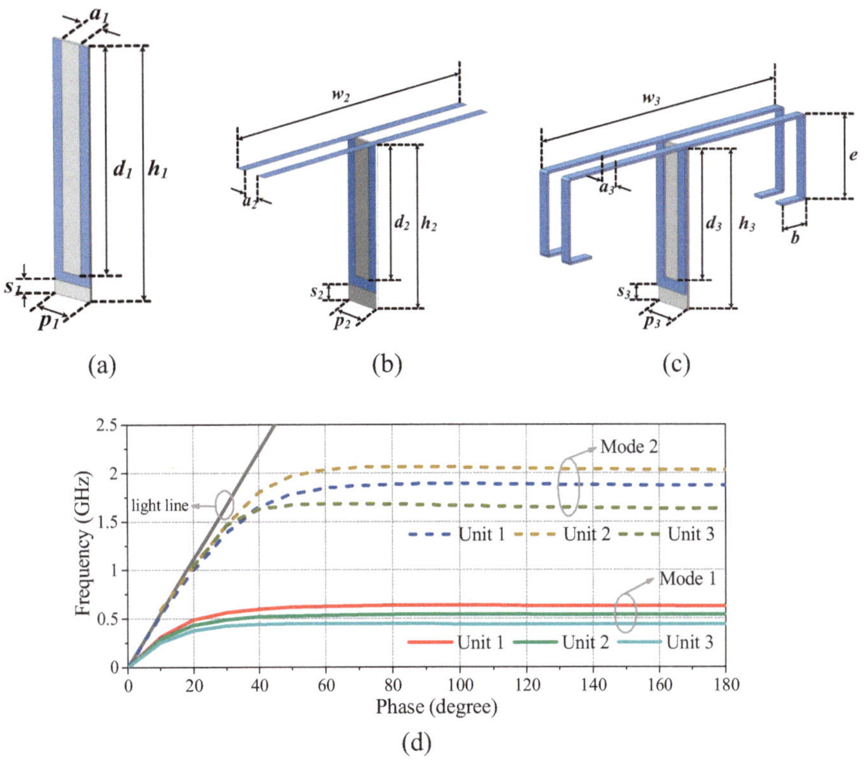

Fig. 7.5 Configuration of three SSPPs unit and it corresponding dispersion curve. **a** unit 1 with h_1 = 105 mm, a_1 = 7.5 mm, d_1 = 95 mm, s_1 = 6 mm; **b** unit 2 with h_2 = 60 mm, a_2 = 7.5 mm, d_2 = 55 mm, s_2 = 6 mm, w_2 = 110 mm; **c** unit 3 with h_3 = 60 mm, a_3 = 7.5 mm, d_3 = 55 mm, s_3 = 6 mm, w_3 = 110 mm, e = 20 mm, b = 10 mm; **d** dispersion curve of unit 1, 2 and 3

Table 7.1 Original parameters of SSPPs Units

Unit 1	h_1	a_1	d_1	p_1			
	105	7.5	95	15			
Unit 2	h_2	a_2	d_2	s_2	w_2		
	60	7.5	55	6	110		
Unit 3	h_3	a_3	d_3	s_3	w_3	e	b
	60	7.5	55	6	110	20	10

*Units: mm

frequencies of the 1st mode and the 2nd mode are 435 MHz and 1.63 GHz respectively. Their working bands are shown in Table 7.2.

C. *Working Modes*

It is obvious that the 1st SSPP mode and the 2nd SSPP mode can work in these three SSPPs structures, respectively. And the working modes maybe mixed modes of odd mode and even mode.

Table 7.2 Working band of SSPPs units

	Mode 1		Mode 2	
	f_s (GHz)	$f_{cut\ off}$ (GHz)	f_s (GHz)	$f_{cut\ off}$ (GHz)
unit 1	0	0.62	0	1.87
unit 2	0	0.54	0.54	2.03
unit 3	0	0.43	0.8	1.63

f_s: start frequency
$f_{cut\ off}$: cut-off frequency

To the unit 1, only the odd SSPP mode travelling wave can run in this kind SSPPs structure, and the current magnitude distribution in 1st mode is close to sinusoidal distribution on one arm, whichjust likes a single pole as depicted in Fig. 7.6a. To the unit 2, the arm folded into T-shape, the current magnitude distribution like sinusoidal too, but diverge into two branches on the top.

To the unit 3, further folded m-shap, at 1st SSPPs mode, the current distribution on the arm is sinusoidal too, as shown in Fig. 7.6b. The current on OB is vertical, and then it is seperated into two parts at point B. At the end points D_1 and D_2 the current is zero equivalent to open circuit. When the unit 3 is extended as SSPPs transmission line and transmitting the traveling wave, the vertical current on OB

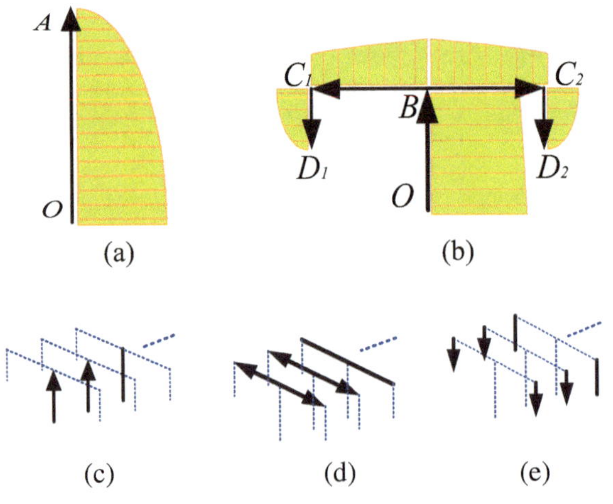

Fig. 7.6 The current magnitude distribution diagram on the arm of the SSPPs unit in 1st SPP mode. **a** Approximated distribution of the current magnitude on one arm of unit 1, which obeys sinusoidal distribution; **b** approximated distribution of the current magnitude on one arm of unit 3; **c** the current direction on the central vertical part of the arm of unit 3, which is odd mode; **d** the current direction on the top horizontal part of the arm of unit 3, which is even mode; **e** the current directions on the end vertical parts of the arm of unit 3, which are odd modes

will be extended to current series like that in Fig. 7.6c, which is a typical odd SSPPs mode. And the horizontal current on BC_1 and BC_2 can be extended to current series like that in Fig. 7.6d, which is a standard even SSPPs mode. Similarly, the current distribution on C_1D_1 and C_2D_2 can be extended to current series like that in Fig. 7.6e, which is odd SSPPs mode, but the current directions are opposite to the current on OB. So, there are mixed SSPPs modes traveling wave running in this case.

D. *Transmission Performances*

The m-shaped SSPPs unit 3 is extended to m-shaped SSPPs transmission line with two ports, and the whole length of the SSPPs transmission lines is 879.5 mm, The electric field distribution at 1.5 GHz is simulated and shown in Fig. 7.7, which shows slow wave and higher mode characteristics. The E_x component distribution on *yoz* plane is in Fig. 7.7b, which shows clear odd mode SSPPs transmission feature. The E_z component distribution is depicted in Fig. 7.7c, which is symmetrically along the y-axis, just like even mode, andwhich is suppressed in the SSPPs wave guide.

Figure 7.7 also explains that the m-shaped SSPPs transmission lines can transmit the mixed mode of E field, which is just corresponding to the current distribution in Fig. 7.6.

The S parameters curves are simulated and given in Fig. 7.8. And the low loss bands are 260–430 MHz and 0.97–1.9 GHz, which are almost identical to the two SSPP working modes in the dispersive curve of unit 3 in Fig. 7.5d.

In addition, the insert loss is 0.3 dB in the first band, and 3.1 dB in the 2nd band, which means the transmission efficiency in the 2nd SSPP transmission mode is smaller than that in the 1st mode. The reason is that the electrical width of the slots

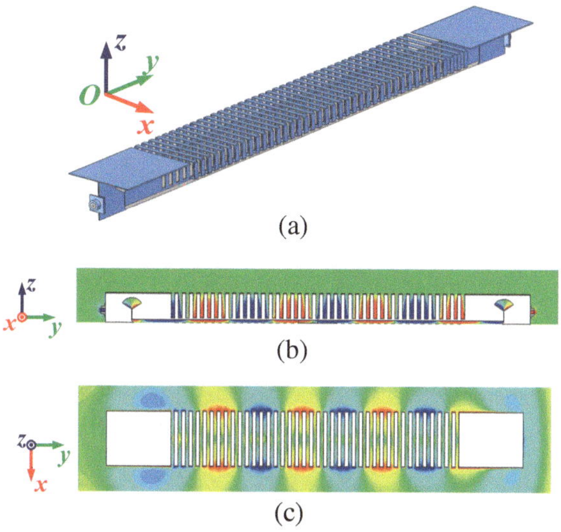

Fig. 7.7 Configuration and E field of SSPPs transmission lines, **a** is the structure of TL 3 based on unit 3; **b** is the Ex component in *yoz* plane at 1.5 GHz (mode 2); **c** is the Ez component in *xoy* plane at 1.5 GHz (mode 2). The total length of TLs are 879.5 m

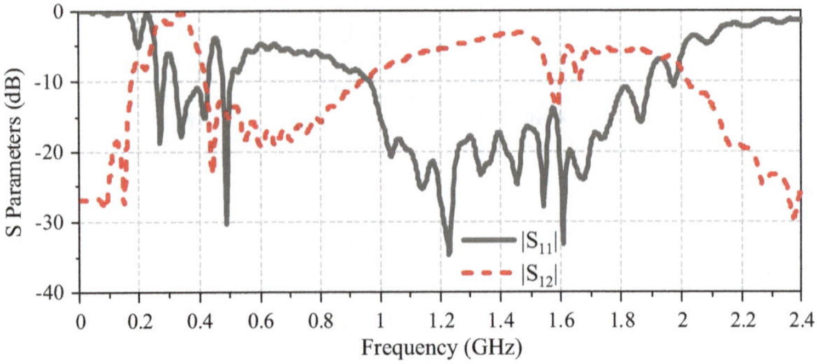

Fig. 7.8 Simulated $|S_{11}|$ and $|S_{12}|$ of transmission line (TL 3) based on m-shaped SSPPs unit 3

in SSPPS unit 3 is larger in higher order frequency band, so that the leakage radiation is stronger from the SSPPs slots than that in the lower frequency band.

7.3.2 Antenna Design

According to the m-shaped SSPPs structure discussed above, a feed network is added at one end, and the other end is improved to be radiative structure. The low profile antenna based on m-shaped SSPPs structure is constructed, as shown in Fig. 7.9. In addition, considering the conformal requirement of the antenna and the car shell, and the requirement of improving its own stability, e = 60 mm is selected, which not only connects the extended end branches with the floor, but also shortens the ground to reduce the size. In Fig. 7.9, yellow is metal, thickened to 1 mm, and blue is FR-4 medium.

The initial parameters of the m-shaped SSPPs antenna are shown in Table 7.3.

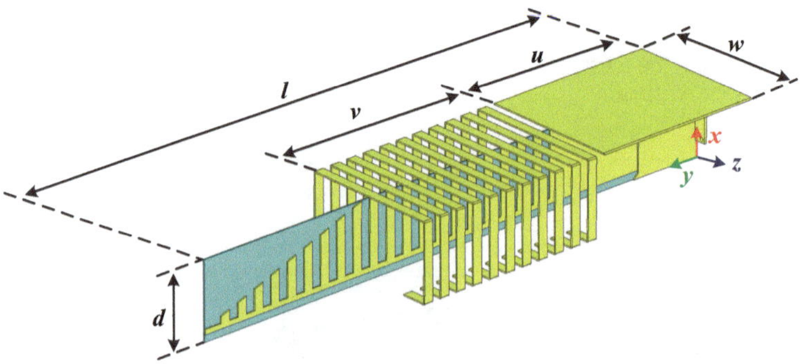

Fig. 7.9 The configuratino of m-shaped SSPPs antenna

Table 7.3 Initial parameter values of m-shaped SSPPs antenna

Par.	a	b	d	h	S	e	l
(mm)	7.5	20	60	50	6	60	450
Par.	l_{f1}	l_{f2}	r_1	r_2	W	w_s	w_{f1}
(mm)	20	50	24	12	110	1	2
Par.	w_{f2}	u	v				
(mm)	1	136	165				

The m-shaped SSPPs antenna need to be placed on the metal surface, and an ideal electric boundary is adopted under the antenna. The S parameters of the antenna obtained by simulation are shown in Fig. 7.10. The operating bandwidth is 447–515 MHz, 0.97–1.81 GHz, which is wide, and it is basically consistent with the passband of m-shaped SSPPs unit.

The simulated gain curve of the m-shaped SSPPs antenna is depicted in Fig. 7.11. The gain is 6–7.5 dB in the low frequency band and 6–12 dB in the high frequency band. The m-shaped SSPPs antenna achieves low profile and high gain performance.

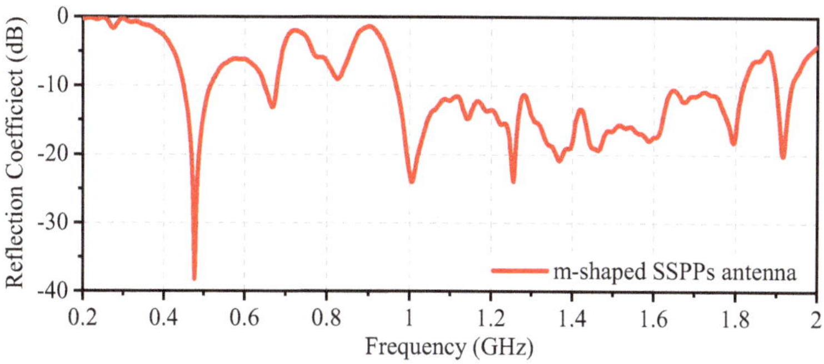

Fig. 7.10 The reflection coefficients of m-shaped SSPPs antenna

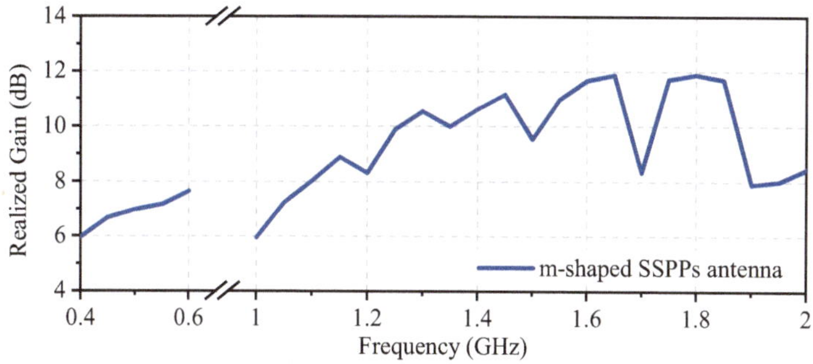

Fig. 7.11 The gain curve of m-shaped SSPPs antenna

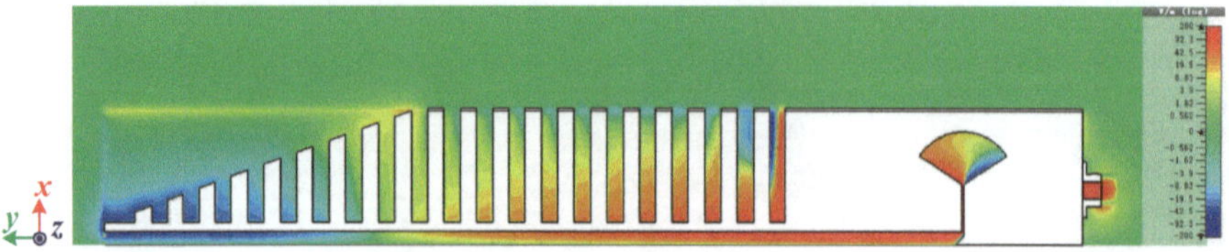

(a) Electric field @ 500 MHz

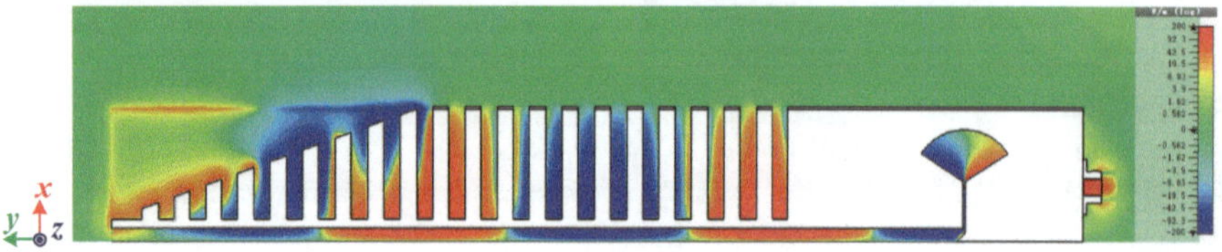

(b) Electric field @ 1.6 GHz

Fig. 7.12 The E_Z component of m-shaped SSPPs antenna at different frequencies

Figure 7.12a shows the simulated electric field's E_Z component distribution at 500 MHz, and Fig. 7.12b shows the electric field E_Z component distribution at 1.6 GHz. It is obvious that the electric field propagates along the m-shaped SSPPs structure and enters the free space at the end.

Figure 7.13 shows the simulated patterns at 500 MHz, 1.4 GHz and 1.6 GHz respectively, which have obvious end fire characteristics. It can be seen that at 500 MHz, the gain of the antenna is 7 dB and the beam width is 138.7°. But the front to back ratio is poor, about 3 dB. At high frequency, the maximum gain of the antenna is 11.6 dB, the beam width is 70° at 1.4 GHz, and the front to back ratio is more than 15 dB.

7.4 Low Profile Spoof Surface Plasmon Polaritons Endfire Antenna and Array

Reference [25] proposed a SSPPs terminal antenna fed by asymmetric gap coupling, as shown in Fig. 7.14. The narrow gap in the structure is used as the feeding balun structure. The experiment shows that the antenna has high gain terminal effect. This section will expand on this basis. On the basis of reference [25], the antenna in Fig. 7.14 is truncated as in Fig. 7.15 because of the actual floor. In order to offset the working frequency of the antenna to low frequency, the antenna is extended and bent at the top. The final simulation structure is shown in Fig. 7.15. The optimized overall antenna is 54.4 mm in height, 497.2 mm in length and 150 mm in width. In order to verify the performance of the antenna, we carried out the processing test, and the physical diagram is shown in Fig. 7.16.

Fig. 7.13 The radiation pattern of m-shaped SSPPs antenna at different frequencies

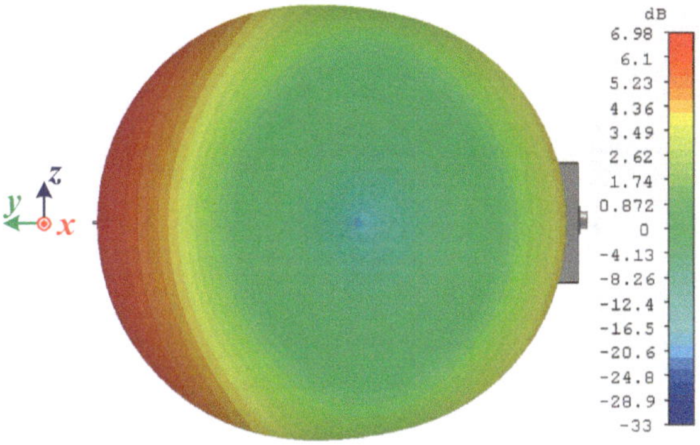

(a) Radiation pattern @ 500 MHz

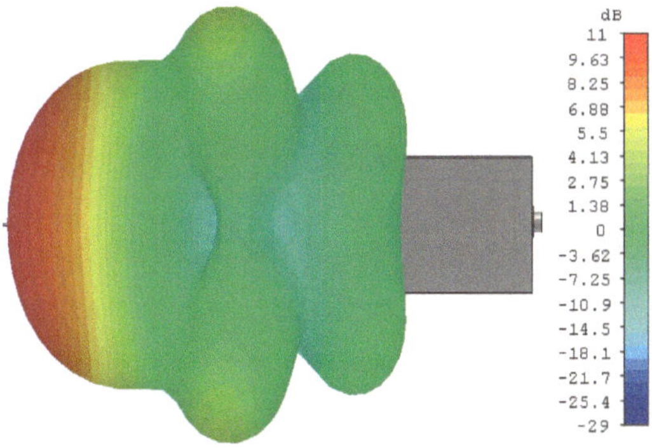

(b) Radiation pattern @ 1.4 GHz

Fig. 7.14 The end-fire antenna based **a** front view; **b** back view

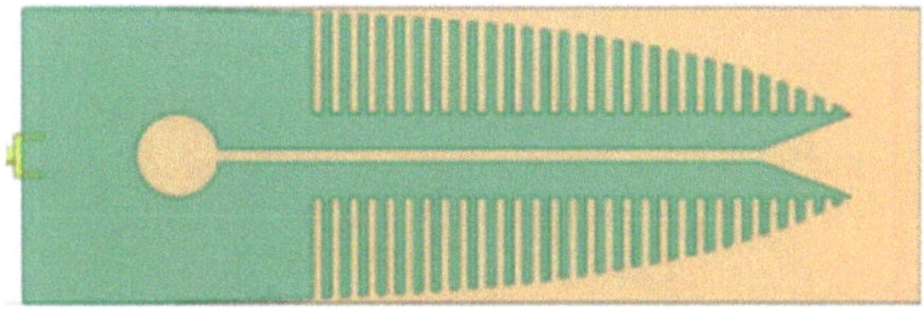

(a)

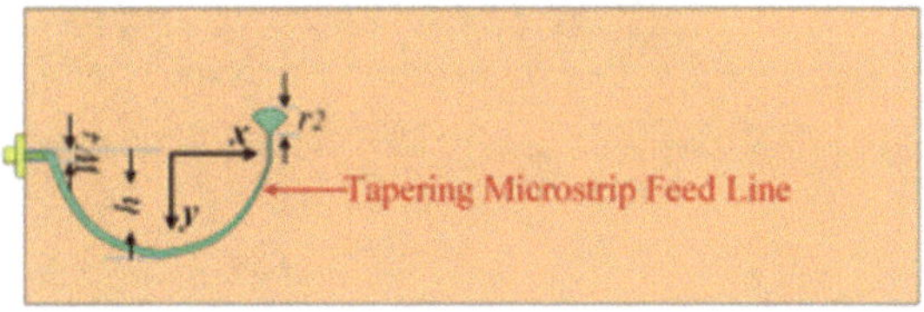

(b)

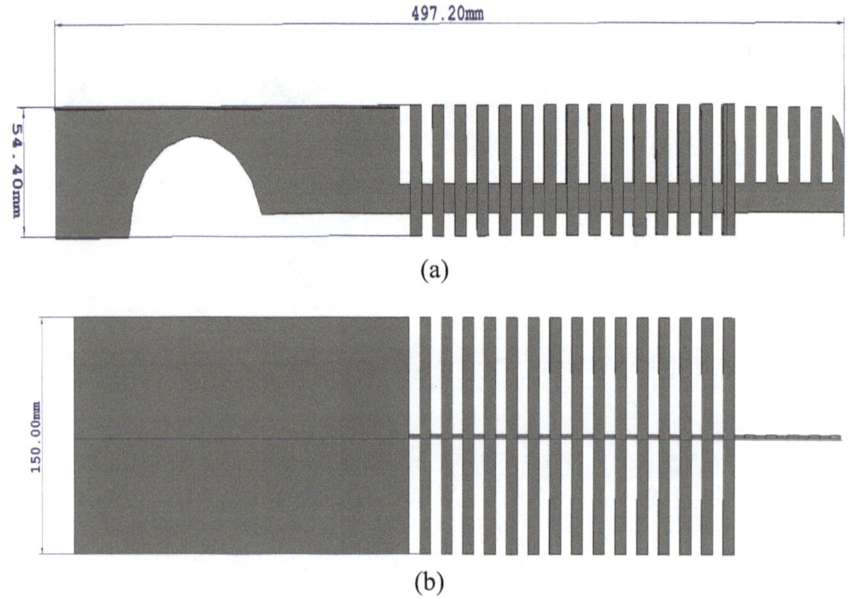

Fig. 7.15 The improved end-fire antenna based **a** front view; **b** vertical view

By processing the data collected by vector network analyzer E8361C, the simulation and measured S parameters and VSWR of the improved low profile end-fire antenna are shown in Fig. 7.17. It can be seen from Fig. 7.17 that the working frequency band of the antenna is 400–500 MHz, and the relative bandwidth is 22.22%. Figure 7.18 shows the measured patterns of the antenna at 400 MHz, 420 MHz and 460 MHz. The results show that the antenna has good endfire effect at 400 MHz, 420 MHz and 460 MHz.

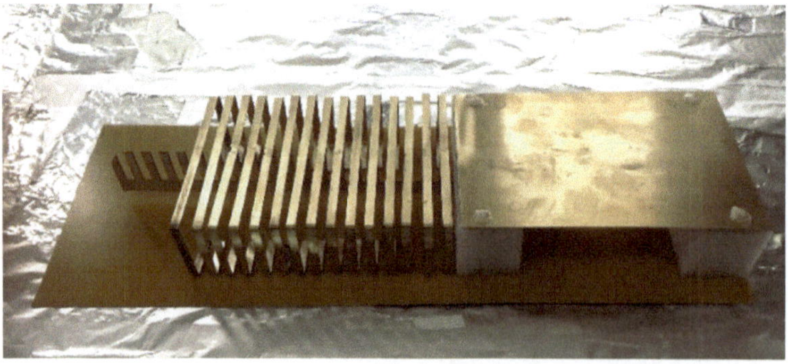

Fig. 7.16 The fabricated prototype of the improved end-fire antenna

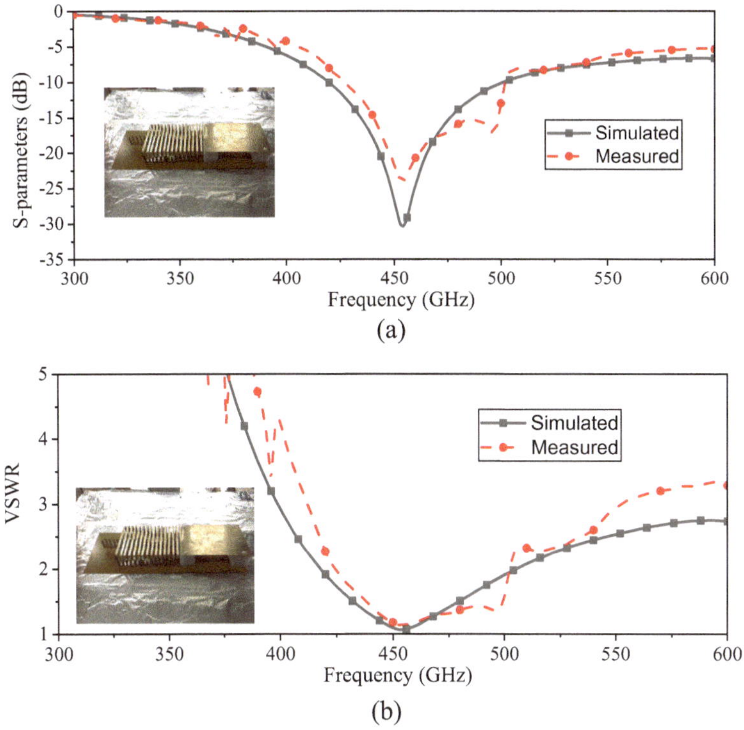

Fig. 7.17 The simulated and measured results of the improved end-fire antenna, **a** S-parameters; **b** VSWR

7.5 Summary of This Chapter

In this chapter, the dispersion, transmission and characteristics of the irregular SSPPs structure isnvestigated. On this basis, m-shaped SSPPs antenna working in UHF band and L-band is proposed. The antenna profile is only 0.1 times of the largest wavelength, and the vertical polarization end fire characteristics are realized. The measured results are in good agreement with the simulation results. The measured results verify the effectiveness of the m-sahped SSPPs antenna, which provides a new idea for the design of vehicle antenna.

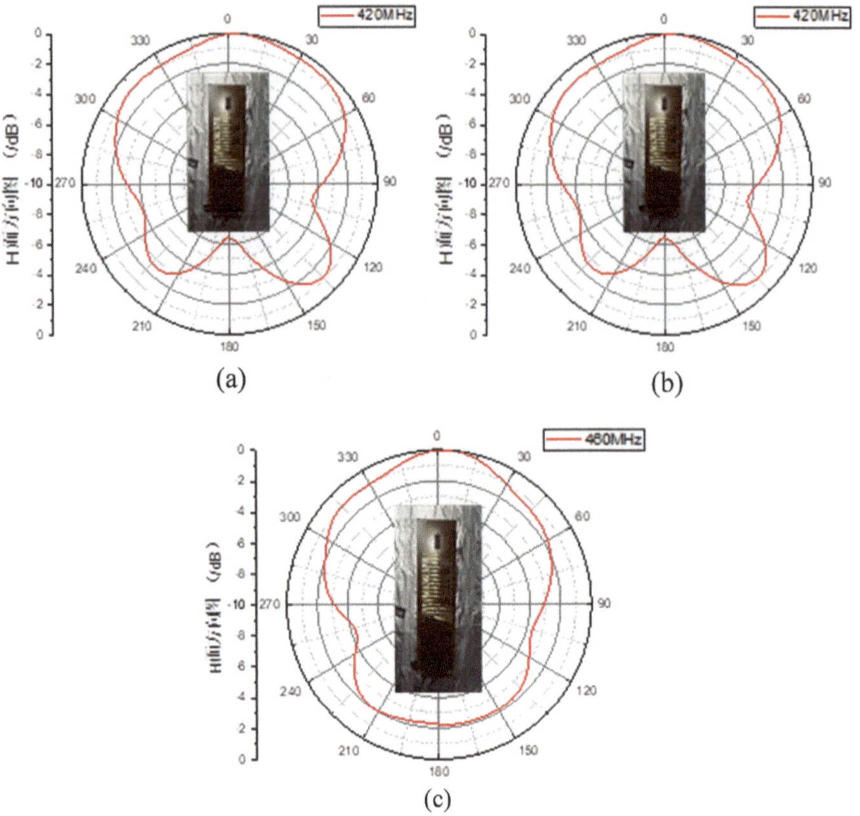

Fig. 7.18 The measured far-filed patterns of the improved end-fire antenna, **a** 400 MHz; **b** 420 MHz; **c** 400 MHz

References

1. K. Li, L. Li, Y. Cai, C. Zhu, C. Liang, A novel design of low-profile dual-band circularly polarized antenna with meta-surface. IEEE Antennas Wirel. Propag. Lett. **14**, 1650–1653 (2015). https://doi.org/10.1109/LAWP.2015.2417169
2. L. Wang et al., Wideband dual-polarized binary coding antenna with wide beamwidth and its array for millimeter-wave applications. IEEE Antennas Wirel. Propag. Lett. **19**(4), 636–640 (2020). https://doi.org/10.1109/LAWP.2020.2974160
3. W.E.I. Liu, et al., Broadband low-profile L-probe fed metasurface antenna with TM leaky wave and TE surface wave resonances. IEEE Trans. Antennas Propag. **68**(3), 1348–1355 (2020)
4. Z.D. Wang, S. Huang, J.L. Liu, Z.L. Lv, X.B. Tang, in *A Low Profile Wideband Linearly Polarized Slot Antenna Using a Circular AMC Surface*. 2020 9th Asia-Pacific Conference on Antennas and Propagation (APCAP), 2020, pp. 1–2. https://doi.org/10.1109/APCAP50217.2020.9246108.

5. H. Wu et al., A low-profile wideband dual-polarized antenna based on an improved HIS and its broad-angle beam-scanning array. IEEE Antennas Wirel. Propag. Lett. **19**(3), 383–387 (2020). https://doi.org/10.1109/LAWP.2020.2967223
6. N.M. Mohamed-Hicho, E. Antonino-Daviu, M. Cabedo-Fabrés, M. Ferrando-Bataller, A novel low-profile high-gain UHF antenna using high-impedance surfaces. IEEE Antennas Wirel. Propag. Lett. **14**, 1014–1017 (2015). https://doi.org/10.1109/LAWP.2015.2389274
7. Z. Chen, Z. Shen, Wideband flush-mounted surface wave antenna of very low profile. IEEE Trans. Antennas Propag. **63**(6), 2430–2438 (2015). https://doi.org/10.1109/TAP.2015.2414948
8. J.B. Pendry, L. Martín-Moreno, F.J. Garcia-Vidal, Mimicking surface plasmons with structured surfaces. Science **305**(5685), 847–848 (2004). https://doi.org/10.1126/science.1098999
9. X. Shen, T. Jun Cui, Planar plasmonic metamaterial on a thin film with nearly zero thickness. Appl. Phys. Lett. **102**(21), 211909–211909 (2013)
10. X. Gao et al., Ultrathin dual-band surface plasmonic polariton waveguide and frequency splitter in microwave frequencies. Appl. Phys. Lett. **102**(15) 2013. https://doi.org/10.1063/1.4802739
11. H.F. Ma, X. Shen, Q. Cheng, W.X. Jiang, T.J. Cui, Broadband and high-efficiency conversion from guided waves to spoof surface plasmon polaritons. Laser Photon. Rev. (2014)
12. X. Shen, T.J. Cui, D. Martin-Cano, F.J. Garcia-Vidal, Conformal surface plasmons propagating on ultrathin and flexible films. Proc. Natl. Acad. Sci. USA **110**(1), 40–45 (2013). https://doi.org/10.1073/pnas.1210417110
13. Y. Yang, X. Shen, P. Zhao, H.C. Zhang, T.J. Cui, Trapping surface plasmon polaritons on ultrathin corrugated metallic strips in microwave frequencies. Opt. Express **23**(6), 7031–7037 (2015). https://doi.org/10.1364/OE.23.007031
14. M. Wang, H.C. Wang, S.C. Tian, H.F. Ma, T.J. Cui, Spatial multi-polarized leaky-wave antenna based on spoof surface plasmon polaritons. IEEE Trans. Antennas Propag. **99**, 1–1 (2020)
15. A. Kianinejad, Z.N. Chen, L. Zhang, W. Liu, C.-W. Qiu, Spoof plasmon-based slow-wave excitation of dielectric resonator antennas. IEEE Trans. Antennas Propag. **64**(6), 2094–2099 (2016). https://doi.org/10.1109/tap.2016.2545738
16. A. Kandwal, Q. Zhang, X.L. Tang, L.W. Liu, G. Zhang, Low-Profile spoof surface plasmon polaritons traveling-wave antenna for endfire radiation, IEEE Antennas Wireless Propag. Lett. **99**, 1–1 (2017)
17. J.Y. Yin et al., Endfire radiations of spoof surface plasmon polaritons. IEEE Antennas Wirel. Propag. Lett. **16**, 597–600 (2017). https://doi.org/10.1109/LAWP.2016.2592512
18. X. Zhang, J. Fan, J. Chen, High gain and high-efficiency millimeter-wave antenna based on spoof surface plasmon polaritons. IEEE Trans. Antennas Propag. **67**(1), 687–691 (2019). https://doi.org/10.1109/TAP.2018.2879847
19. J.J. Wu et al., High-directivity radiation based on the leaky mode of spoof surface plasmon polaritons. Microwaves Antennas Propag. Iet **8**(13), 1075–1079 (2014)
20. K. Wang et al., in *Pattern Reconfigurable Antenna Applying Spoof Surface Plasmon Polaritons*, 2019 13th European Conference on Antennas and Propagation (EuCAP), 31 March-5 April 2019, pp. 1–2 (2019)
21. K. Zhuang et al., Pattern reconfigurable antenna applying spoof surface plasmon polaritons for wide angle beam steering. IEEE Access **7**, 15444–15451 (2019). https://doi.org/10.1109/access.2019.2895106
22. H. Yajuan, Z. Jieqiu, L. Yongfeng, W. Jiafu, Q. Shaobo, Z. Anxue, 360 scanning multi-beam antenna based on spoof surface plasmon polaritons. Acta Physica Sinica **65**(14) (2016)
23. Y. Han et al., Multibeam antennas based on spoof surface plasmon polaritons mode coupling. IEEE Trans. Antennas Propag. **65**(3), 1187–1192 (2017). https://doi.org/10.1109/TAP.2016.2647588

24. Y. Fan et al., Frequency scanning radiation by decoupling spoof surface plasmon polaritons via phase gradient metasurface. IEEE Trans. Antennas Propag. **66**(1), 203–208 (2018). https://doi.org/10.1109/TAP.2017.2767625
25. K. Zhuang, J.P. Geng, Z. Ding et al., A compact endfire radiation antenna based on spoof surface plasmon polaritons in wide bandwidth. Progress Electromag. Res. **79**, 147–157 (2019)
26. X Cheng, J. Geng, C. Ren, H. Zhou, X. Zhao, K. Wang, J. Han, X. Liang, R. Jin. In *Low profile Vertically Polarized Endfire Antenna Based on Spoof Surface Plasmon Polaritons*. Proceedings of IEEE AP-S Society of International Symposium, July 2020, pp. 299–230. 2020 IEEE International Symposium on Antennas and Propagation and North American Radio Science Meeting (AP-S/URSI 2020, Montréal, Québec, Canada (2020)
27. G.S. Kong, W. Meng, H.F. Ma, in *Low Sidelobe Leaky-Wave Antenna Based on Spoof Plasmonic Waveguide*, 2016 Progress in Electromagnetic Research Symposium (PIERS), 8–11 Aug 2016, pp. 386–389 (2016). https://doi.org/10.1109/PIERS.2016.7734344
28. G. Kong, H. Ma, in *Omnidirectional Antenna Based on Modulated Spoof Surface Plasmon Polaritons Waveguide*, 2015 Asia-Pacific Microwave Conference (APMC), 6–9 Dec 2015, vol. 3, pp. 1–3 (2015). https://doi.org/10.1109/APMC.2015.7413325

Chapter 8
Pattern Reconfigurable Antenna Based on Spoof Surface Plasmon Polaritons

Junping Geng, Kun Wang, and Kaijie Zhuang

Compared with the traditional antenna, the reconfigurable antenna is easy to achieve diversity, can greatly expand the working range and capacity of the antenna, and is small in size, easy to integrate with the control circuit [1, 2]. Reconfigurable antenna mainly includes frequency reconfigurable antenna, pattern reconfigurable antenna and polarization reconfigurable antenna. Especially, the pattern reconfigurable antenna can improve the effective coverage area of antenna with high gain and avoid noise interference.

Phased array is the most typical method of pattern reconfiguration, but it is difficult to achieve ultra wide angle scanning pattern reconfiguration because of its large volume. The mechanical scanning antenna is also a pattern reconfigurable antenna, which can realize pattern switching at any angle, but its structure is complex and its volume is huge. Leaky wave antenna is also a kind of pattern reconstruction antenna, but it is mainly the change of beam with frequency [3, 4], and it is difficult to achieve large angle switching of pattern at the same frequency. In recent years, a variety of SSPPs antennas have been proposed, which can work in different modes, including broadside radiation, end fire radiation and even higher-order mode.

In this chapter, based on the study of leaky wave antenna and end fire antenna based on SSPPs structure, the idea of broadside radiation and end fire radiation reconfiguration is proposed. Through the reconfigurable circuit constructed by PIN diode, the switching between odd mode and even mode of SSPPs antenna is realized, and the purpose of broadside pattern and end fire pattern reconfiguration n is achieved [5].

J. Geng (✉) · K. Wang · K. Zhuang
Department of Electronic Engineering, Shanghai Jiao Tong University, Shanghai, China
e-mail: gengjunp@sjtu.edu.cn

K. Wang
e-mail: wanggkun@sjtu.edu.cn

J. Geng et al., *Spoof Surface Plasmon Polaritons Antenna*, Modern Antenna,
https://doi.org/10.1007/978-981-16-4721-5_8

8.1 Basic Principle of Pattern Reconfigurable Antenna Based on Spoof Surface Plasmon Polaritons

8.1.1 Endfire Antenna Based on Spoof Surface Plasmon Polaritons

Figure 8.1 shows the structure of the monopole-fed end-fire antenna based on the artificial surface plasmon. The two ends of the SSPPs transmission line are tapered. The electromagnetic energy of the monopole can be coupled into the SSPPs transmission line by the left tapered structure and radiated out in the end tapered structure. Figure 8.2 shows the impedance matching curve of the antenna. Because of the resonant characteristics of the feed, the impedance bandwidth of the SSPPs terminal antenna is very narrow. Figure 8.3 show the far-field pattern of the antenna and the z-component distribution of the electric field. It can be seen that odd modes are mainly transmitted in SSPPs transmission lines.

8.1.2 Broadside Antenna Based on Spoof Surface Plasmon Polaritons

Electromagnetic waves transmitted in SSPPs do not radiate directly into free space because the number of waves does not match that in free space. In order to convert SSPPs mode into free space radiation mode, periodic modulation units are used in [6–8]. Among them, the paper [6] introduces a leaky wave antenna based on SSPPs. In this section, a simple leaky-wave antenna is designed and simulated based on the model in Ref. [6] to analyze the transmission and radiation characteristics of this antenna. Figure 8.4 show the basic model of this antenna. The characteristic impedance of coplanar waveguide is designed as 50 Ω; Both ends of SSPPs transmission line adopt a gradual structure, and its principle is consistent with that described in Sect. 8.1.1; Four-cycle rhombic metal patches are placed directly on SSPPs transmission lines to convert electromagnetic modes in SSPPs transmission lines into radiation modes in free space.

Figure 8.5 show the impedance matching curve of this antenna. As can be seen, the antenna in 5–6.5 GHz $|S_{11}|$ less than −10 dB. Figure 8.6 shows the pattern of

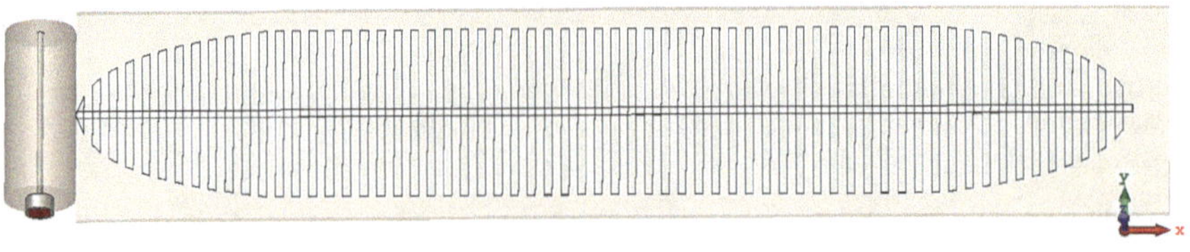

Fig. 8.1 SSPPs based endfire antenna fed by a monopole

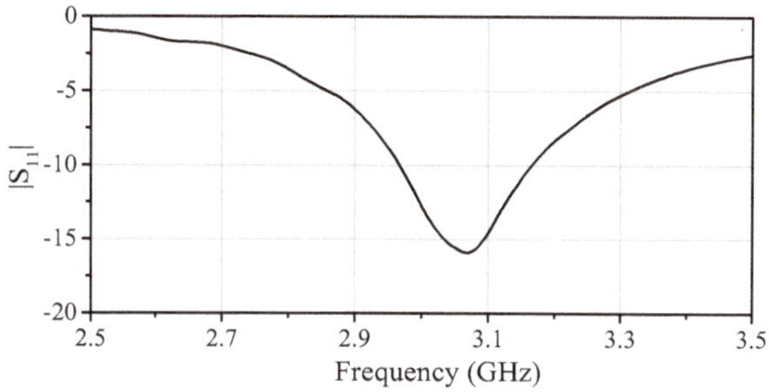

Fig. 8.2 The $|S_{11}|$ performance of the proposed SSPPs based endfire antenna

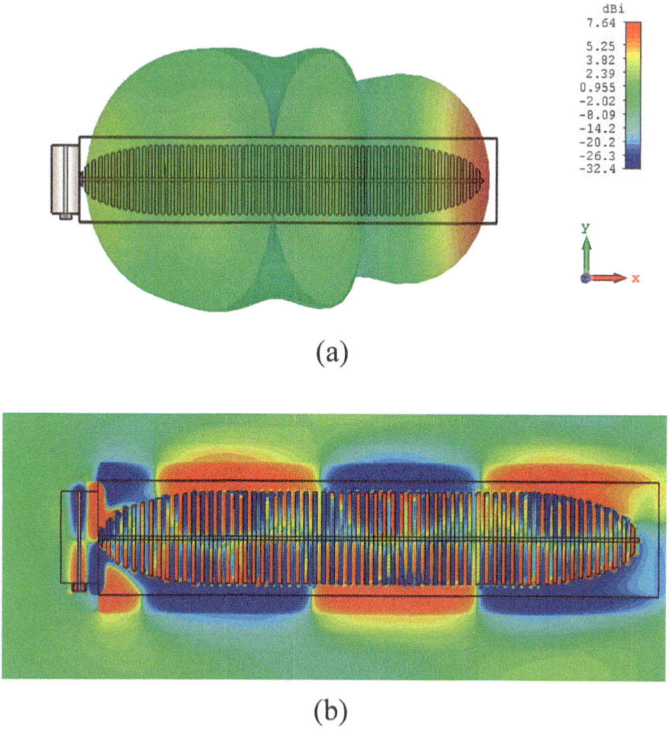

Fig. 8.3 SSPPs based endfire antenna. **a** and **b** are radiation pattern and electric fields distribution (z component) at 3.1 GHz

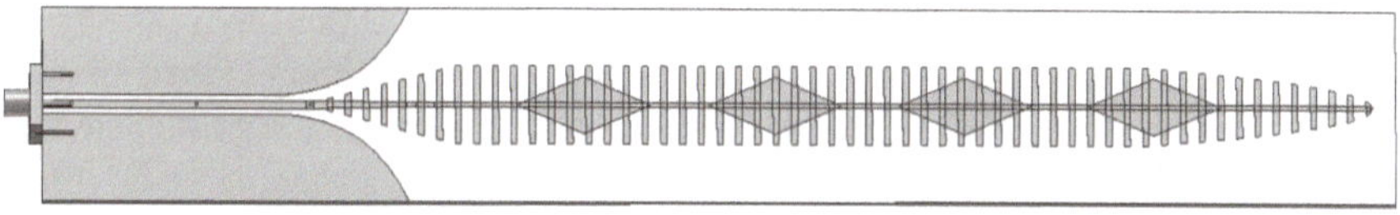

Fig. 8.4 The leaky wave antenna based on SSPPs

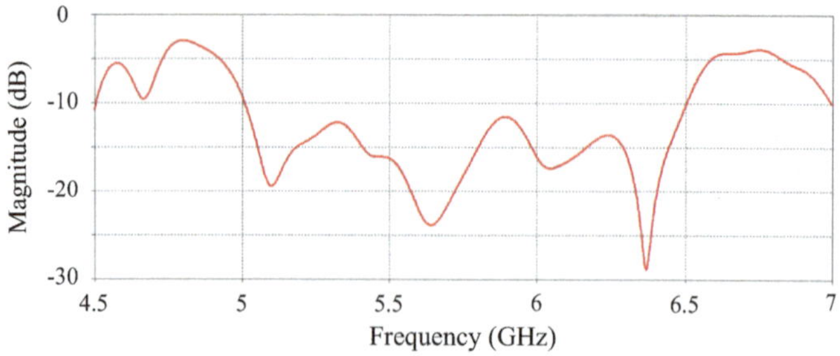

Fig. 8.5 The $|S_{11}|$ performance of the proposed leaky wave antenna based on SSPPs

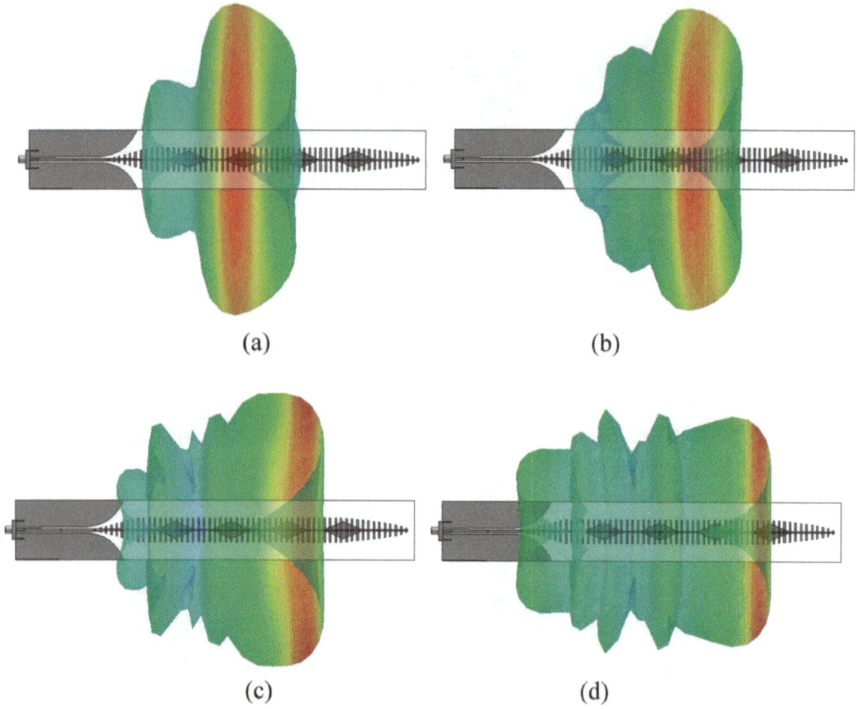

Fig. 8.6 Radiation pattern at 5.2 GHz, 5.4 GHz, 5.6 GHz, and 5.8 GHz, respectively

the antenna at 5.2, 5.4, 5.6 and 5.8 GHz. It can be seen that this SSPPs-based leaky-wave antenna is side-beam, and with the increase of frequency, the beam direction of side-beam gradually tends to the end-beam direction. When the frequency increases to 5.8 GHz, the side lobe becomes higher and the gain decreases. Figure 8.7 is the near-field electric field distribution of the antenna at 5.5 GHz.

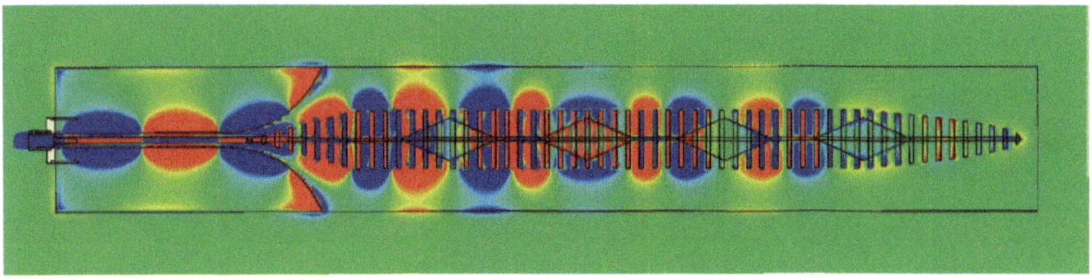

Fig. 8.7 Electric fields distribution (z component) at 5.5 GHz

Compared with Fig. 8.3b, it can be seen that the near-field electric field of the antenna is symmetrical distribution, that is to say, the transmission of the SSPPs transmission line is mainly dual mode.

8.1.3 Principle Verification

Sections 8.1.1 and 8.1.2 discuss an end-fire antenna and a side-fire leaky-wave antenna based on SSPPs structure respectively, and discuss their radiation pattern and transmission mode. It can be seen that there are two basic transmission modes in the traditional SSPPs transmission lines—odd mode and even mode. Among them, when the antenna radiates the end beam, the transmission in SSPPs transmission line is mainly odd mode; When the antenna pattern is lateral, the main transmission mode in SSPPs transmission line is even mode. Therefore, it is envisaged to switch the main transmission mode in SSPPs transmission line to realize the switching of antenna pattern in two states of end-fire and side-fire. Based on the Sects. 8.1.1 and 8.1.2, a simple antenna is designed for verification, as shown in Fig. 8.8.

The antenna parameters shown in Fig. 8.8 are completely consistent with those shown in Fig. 8.4. The difference is that the antenna shown in Fig. 8.8 introduces an additional metal branch to destroy the original symmetric structure. Figure 8.9 shows the electric field distribution of the antenna at 5 GHz. Compared with Fig. 8.7, it can be seen that the original symmetric electric field is destroyed and the odd mode becomes the main component.

Figure 8.10 shows the antenna pattern at 5 and 5.4 GHz. It can be seen that the main beam of the antenna switches from the side-fire mode shown in Fig. 8.6 to the

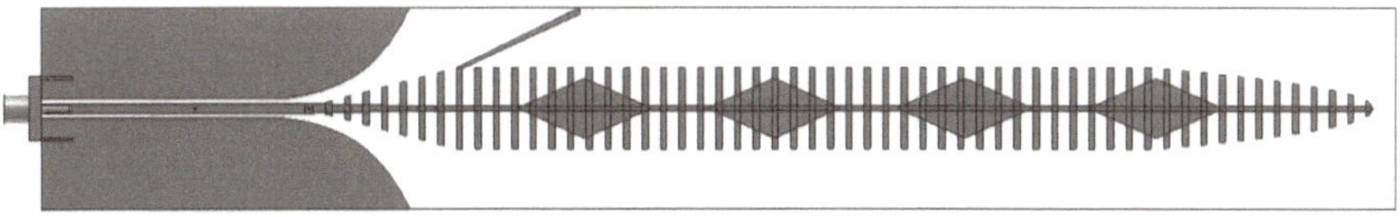

Fig. 8.8 SSPPs antenna with additional circuit

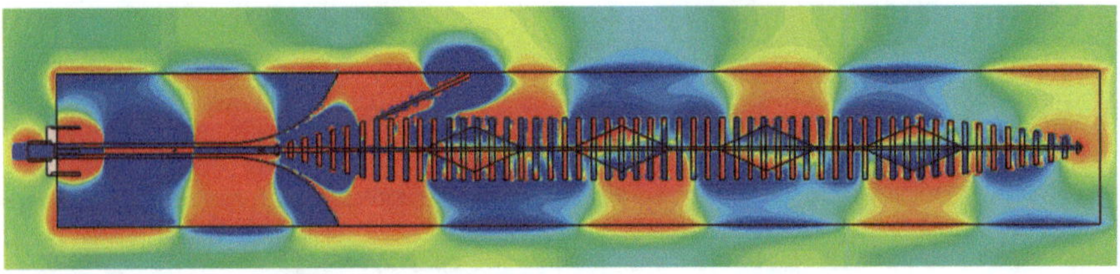

Fig. 8.9 Electric fields distribution at 5 GHz (z component)

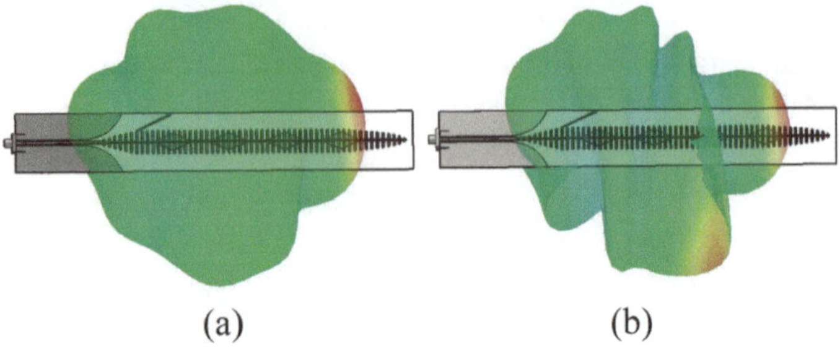

Fig. 8.10 Radiation pattern at **a** 5 GHz; **b** 5.4 GHz

end-fire mode at 5 and 5.4 GHz. However, since the additional branch cannot switch the even mode into odd mode in a wide frequency band, when the frequency increases to 5.4 GHz, the side lobe of the antenna gradually increases.

This section combines the transmission and radiation characteristics of the SSPPs end-fire antenna and side-fire antenna in Sects. 8.1.1 and 8.1.2, on the basis of the side-fire leaky wave antenna in 8.1.2, an additional metal branch is introduced, so that the even mode mode in the SSPPs transmission line is switched to the odd mode mode. Finally, the antenna pattern is given, and the far-field results are consistent with the expected: this kind of SSPPs antenna shows side-fire characteristics in the even mode, and will show end-fire characteristics in the odd mode.

8.2 Pattern Reconfigurable Antenna Applying Spoof Surface Plasmon Polaritons

8.2.1 Antenna Design and Analysis

Two fundamental SSPP modes, odd mode and even mode, are supported on conventional SSPPs transmission line as discussed in [9] and [10]. The EM wave supported in traditional double-side corrugated SSPPs structure can be converted into broadside radiation beam in the even mode as reported in [11]. On the other hand, the endre antenna with similar SSPPs structure has been reported in [12] which shows the characteristic of the odd mode. It means by using certain mode convertor, an antenna with same structure could exhibit totally different radiation characteristic. Based on this law, a radiation pattern recongurable antenna is proposed in this paper.

A. *Antenna Structure*

The framework diagram of the proposed antenna is shown in Fig. 8.11. A reconfigurable circuit is inserted between the CPWand the SSPPs structure to reconfigure the SSPPs mode. By switching between the odd mode and the even mode on the SSPP structure, the radiation pattern can be reconfigured easily.

The configuration of the proposed SSPPs pattern reconfigurable antenna is shown in Fig. 8.12a. The proposed antenna contains four parts, a coplanar waveguide, a transition part, a modulated SSPPs transmission line [11] and a tapering metallic structure. The entire metal structure is fabricated on a 0.762-mm-thick Arlon AD600 substrate (with "$\varepsilon_r = 6.2$ and $\tan\sigma = 0.003$"). The CPW is designed to match the impedance of 50 Ω. The modulated SSPPs transmission line is a corrugated metallic strip periodically decorated with rhombic patches as shown in Fig. 8.12a. The rhombic patches are periodically loaded to improve the performance in broadside radiation state [12]. The diagonal lengths of the rhombic patches are 21.4 mm and 9.6 mm. The height h is about $\lambda_c/4$ (λ_c being the wavelength at center frequency). Figure 8.12b shows the DC bias of the proposed antenna. Metal-via-hole I is connected with DC ground. Two inductors of 220 nH are introduced to minimize the impact of additional DC bias on the antenna.

B. *Reconfigurable Circuit and Excitation of SPP Mode*

The details of the antenna reconfigurable circuit are shown in Fig. 8.13a. Three pin diodes (MA4AGFCP910, 4.2 Ω conduction resistance under positive bias and 0.02 pF capacitance under negative bias) S1, S2 and S3 constitute a reconfigurable control circuit, and S1, S2 and S3 are placed anticlockwise. The simplified circuit and current distribution in even mode state are shown in Fig. 8.13b, and the simplified circuit and current distribution in odd mode state are shown in Fig. 8.13c. It can be seen that when S1 is on and S2 and S3 are off, the current on SSPPs transmission line distributes symmetrically and is in even mode transmission state. At this time, the electric field along y direction is added up and down to offset, just forming a pit in the end fire direction (x direction); when S1 is off and S2 and S3 are

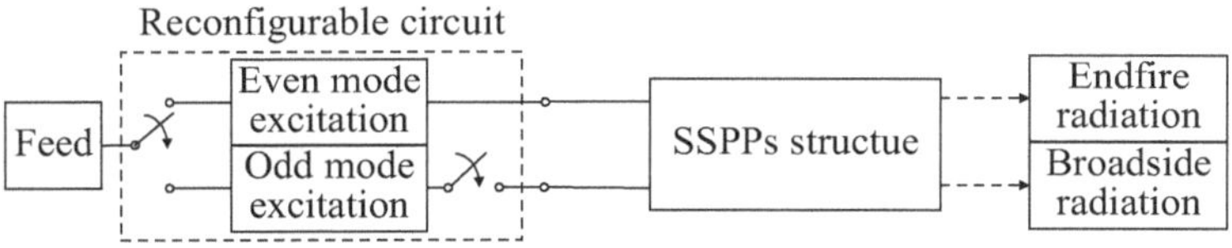

Fig. 8.11 The framework diagram of the proposed antenna. Figure reproduced with permission from Ref. [5], © 2019 IEEE

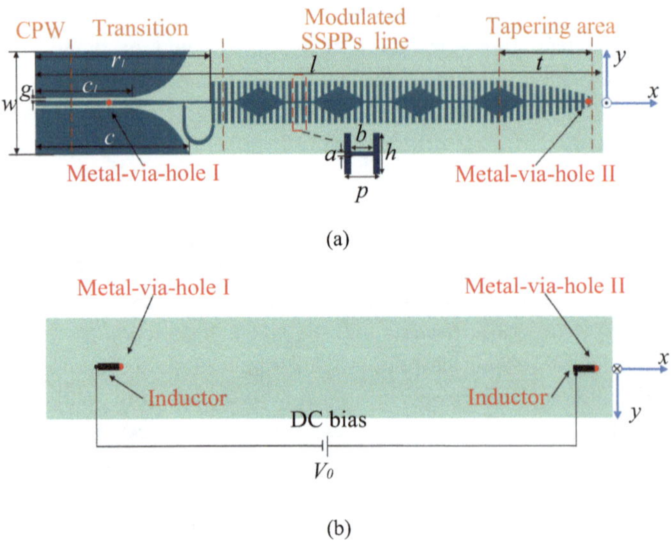

Fig. 8.12 Proposed pattern reconfigurable antenna. **a** Front view; **b** back view of the proposed antenna

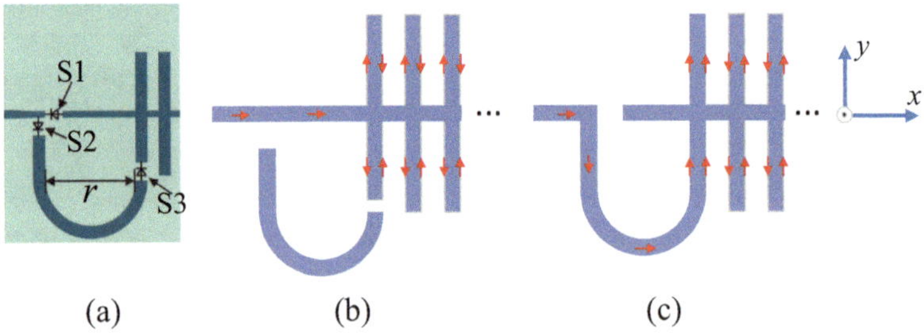

Fig. 8.13 **a** Reconfigurable part of the proposed antenna; simplified current distribution diagrams for **b** even mode and **c** odd mode

on, the current on SSPPs transmission line distributes antisymmetrically and is in odd mode transmission state, and the electric field along y direction is in odd mode transmission state Add up and down in the same direction, just in the end fire direction (x direction) to form a prominent beam.

Figure 8.14a shows the simulated electric field Z component in even mode, corresponding to S1 on, S2 and S3 off states. Figure 8.14b shows the Z component of the electric field during odd mode transmission, corresponding to two states of S1 cutoff, S2 and S3 conduction. In the specific simulation, the lumped port value given by the data sheet of its equivalent parameters is used to replace the pin diode.

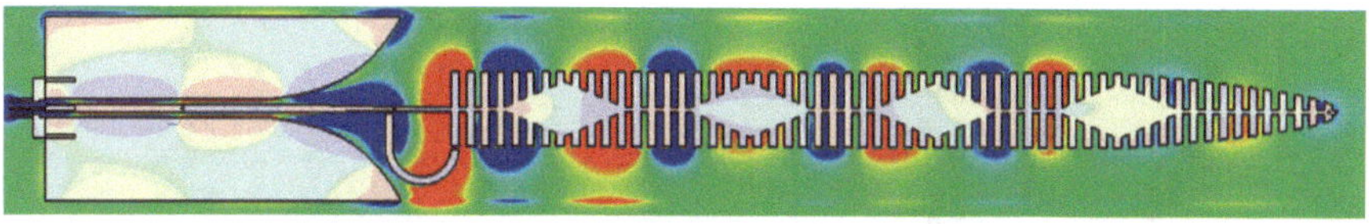

(a)

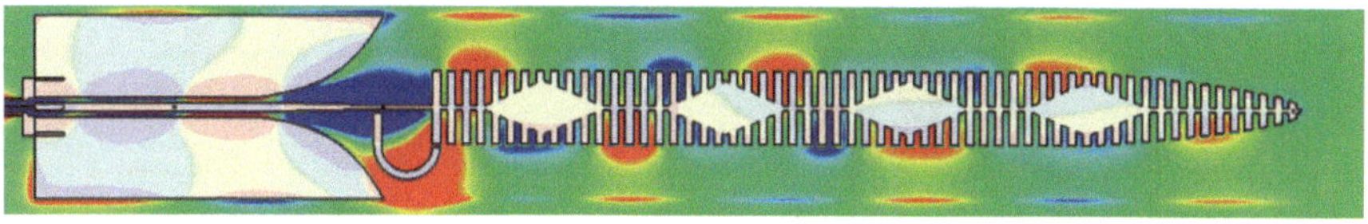

(b)

Fig. 8.14 **a** The simulated results of electric fields distributions (z component) in the even mode; **b** the simulated results of electric fields distributions (z component) in the odd mode

The simulation results in Fig. 8.14 also show that the combination of three pin diodes can reconstruct the transmission mode on SSPPs.

At 5.5 GHz, the power flow distribution of the antenna in the even mode state is shown in Fig. 8.15a. As the electromagnetic wave propagates in the SSPPs transmission line, the electromagnetic energy is continuously radiated to both sides. The power flow distribution of the antenna in odd mode at 5.5 GHz is shown in Fig. 8.15b. It can be seen that the electromagnetic energy in odd mode is well bound in the SSPPs transmission line and transmitted to the end. The corresponding antenna pattern in the even mode state is shown in Fig. 8.16a, which has obvious side radiation characteristics. The antenna pattern corresponding to the odd mode is shown in Fig. 8.16b, and the main beam is obviously in the end fire radiation. It can be seen from the power flow distribution of Fig. 8.15 and the far-field pattern of Fig. 8.16 that when the SSPPs antenna works in even mode, the far-field of the antenna presents the side radiation character, while when the SSPPs antenna works in odd mode, the far-field of the antenna presents the end fire character.

C. *Key Parameter Analysis*

In the reconfigurable antenna, the gradual structure of the end has a great influence on the radiation characteristics of the antenna, and the curvature of the reconfigurable part is the diameter *r* of the microstrip line, which has a great influence on the impedance matching of the reconstructed state.

Figure 8.17 shows the antenna gain versus frequency curves of odd mode and even mode for SSPPs reconfigurable antenna with or without the tapered end structure. It can be seen that the antenna gain can be improved by introducing the tapered structure at the end, especially in the end fire radiation, the tapered structure at the end is an important part of the radiation structure, which has a great impact on the antenna end fire gain.

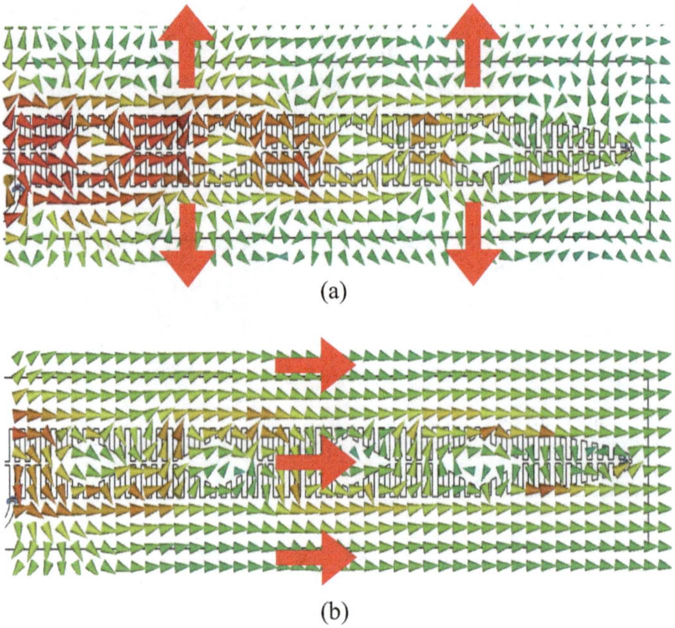

Fig. 8.15 Power flow distribution at 5.5 GHz. **a** The even mode; **b** the odd mode

Fig. 8.16 Far field radiation pattern at 5.5 GHz in **a** the even mode, **b** the odd mode

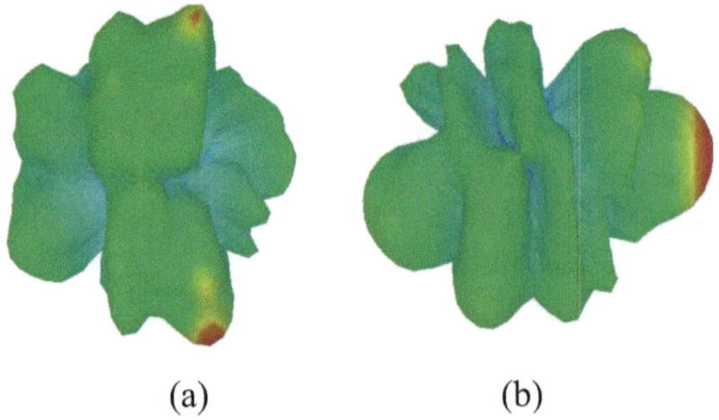

r is an important parameter of the reconfigurable circuit, which mainly affects the conversion efficiency of electromagnetic wave from CPW mode to SSPPs mode. Figure 8.18 shows the change of antenna return loss curve in odd mode and even mode when R changes. When *r* is increased to 10.8 mm, the impedance matching of end emitter and side emitter in the frequency band above 5.4 GHz can be improved, but the low-frequency impedance matching of antenna in side emitter mode will be deteriorated.

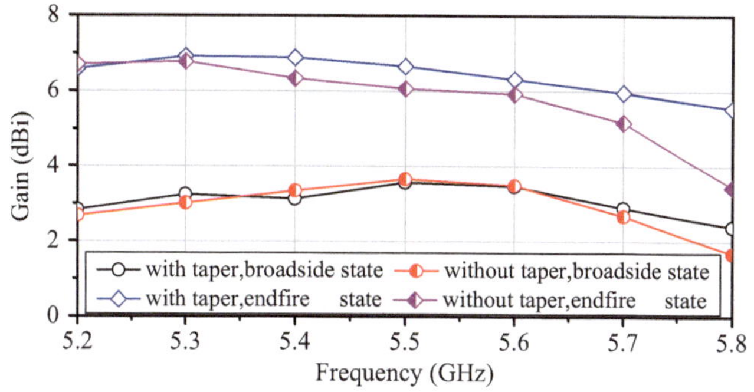

Fig. 8.17 Comparison of simulated gains with and without the tapering

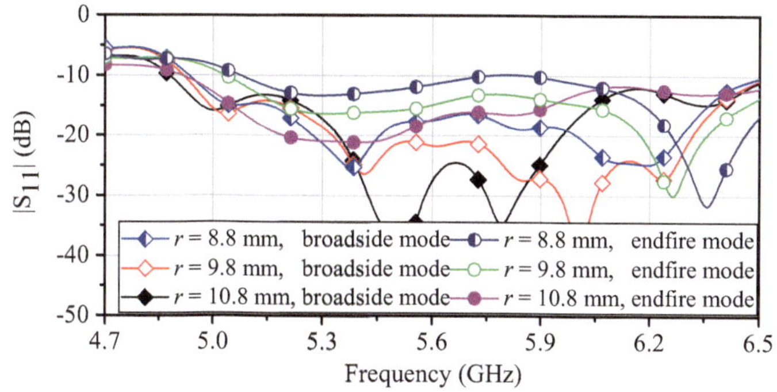

Fig. 8.18 Effect of r on the $|S_{11}|$ performance

8.2.2 Experimental Result and Discussion

After simulation and optimization, the optimal parameters of the antenna are obtained, $a = 0.7$ mm, $b = 1.2$ mm, $c = 59.7$ mm, $g = 0.18$ mm, $h = 12.6$ mm, $l = 218.6$ mm, $p = 2.5$ mm, $r_1 = 68.1$ mm, $r = 9.8$ mm, $t = 35.6$ mm, $w = 31.7$ mm. On this basis, the antenna prototype is processed, as shown in Fig. 8.19, and the far-field pattern is measured in the Satimo chamber, as shown in Fig. 8.20.

Figure 8.19 shows the processed antenna. The blue thin line is the DC bias line. The diameter of the blue thin line should be as thin as possible to reduce the influence of DC bias on the antenna. Figure 8.20 is the test diagram of the antenna in the chamber. The bias voltage and bias current of even mode and odd mode are shown in the lower left corner and upper right corner respectively. In the experiment, 1.4 V is selected as the conduction voltage of a single diode, and the bias current is about 10 mA.

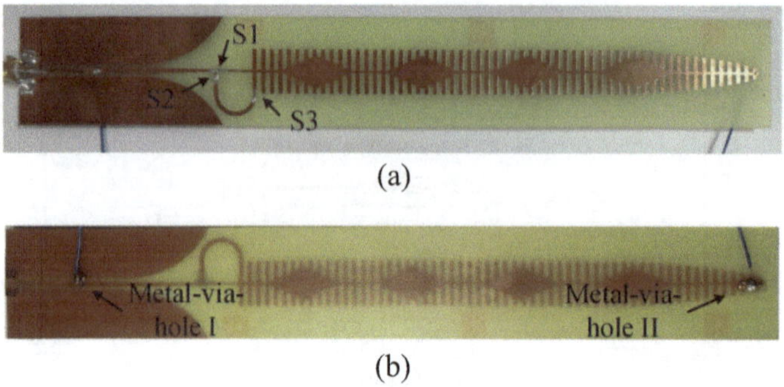

Fig. 8.19 Fabricated prototype of the proposed antenna. **a** Front view and **b** back view

Fig. 8.20 Test environment in SATIMO

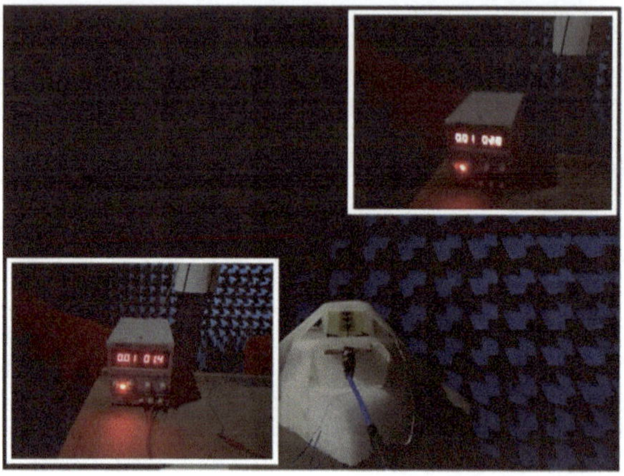

Figure 8.21 shows the return loss curve of the antenna. It can be seen that the $|S_{11}| < -10$ dB frequency band of the antenna in broadside radiation is 4.9–6.5 GHz, and the absolute bandwidth is about 1.6 GHz. In the end fire, the $|S_{11}| <$ -10 dB band of the antenna is 4.95–6.5 GHz, and the absolute bandwidth is about 1.5 GHz. The measured $|S_{11}|$ curve of the antenna is basically consistent with the simulated $|S_{11}|$ curve, but the center frequency point is offset.

The measured and simulated normalized patterns of the antenna at 5.3, 5.5 and 5.7 GHz are shown in Fig. 8.22. Figure 8.22a, c and e are the end fire patterns. It can be seen that the main beam direction of the end fire patterns corresponding to different frequencies basically remains unchanged, pointing to −2°, 2° and 4° respectively. The main beam width basically remains same, and the side lobe is asymmetric, mainly due to the asymmetric effect of the reconfigurable circuit. Figure 8.22b, d and f are the patterns in the side radiation state. It can be seen that the beam directions of the antenna gradually deviates from the side beam direction with the increase of frequency. The beam direction of the antenna is 84° at 5.3 GHz,

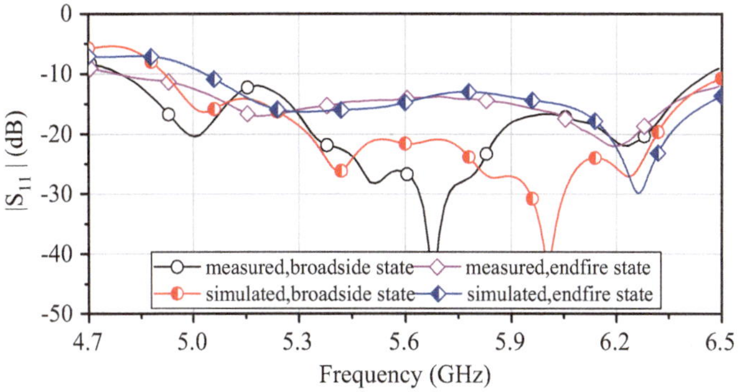

Fig. 8.21 Comparison of the simulated and measured $|S_{11}|$ values of the proposed pattern reconfigurable antenna

68° at 5.5 GHz and 62° at 5.7 GHz. Similarly, because the reconfigurable circuit is partially asymmetric, the antenna pattern is not asymmetric.

The measured beam direction and gain of the antenna are given in Table 8.1.

8.2.3 Results Analysis and Improvement

Although the reconfigurable antenna designed in the previous section produces a good end fire in odd mode, and the antenna works in side radiation in even mode. However, it can be seen from Table 8.1 that the gain of the antenna is not very high when it is in side radiation. In the next, we propose specific some improvement measures to improve the radiation performance of the antenna in side radiation.

The SSPPs antenna operating in side radiation mode is essentially a periodically modulated leaky wave antenna. In this case, increasing the length of SSPPs structure and the number of rhombic patches can improve the antenna aperture and gain. The schematic diagram of antenna aperture in side radiation mode is shown in Fig. 8.23. Figure 8.24 shows the comparison of the radiation patterns of the antenna in the side radiation mode under different antenna apertures. Among them, Fig. 8.24a is the side radiation pattern of the SSPPs antenna model designed in the last section at 5.5 GHz, the length of antenna is 218.7 mm, and there are four diamond patches in the middle; Fig. 8.24b is that the length of SSPPs antenna increases to 341.7 mm, and the number of diamond patches increases to 8. The other parameters remain unchanged.

It can be seen that the gain of Fig. 8.24b is 4.53 dB, which indicates that the antenna gain can be significantly improved by increasing the radiation aperture of the antenna.

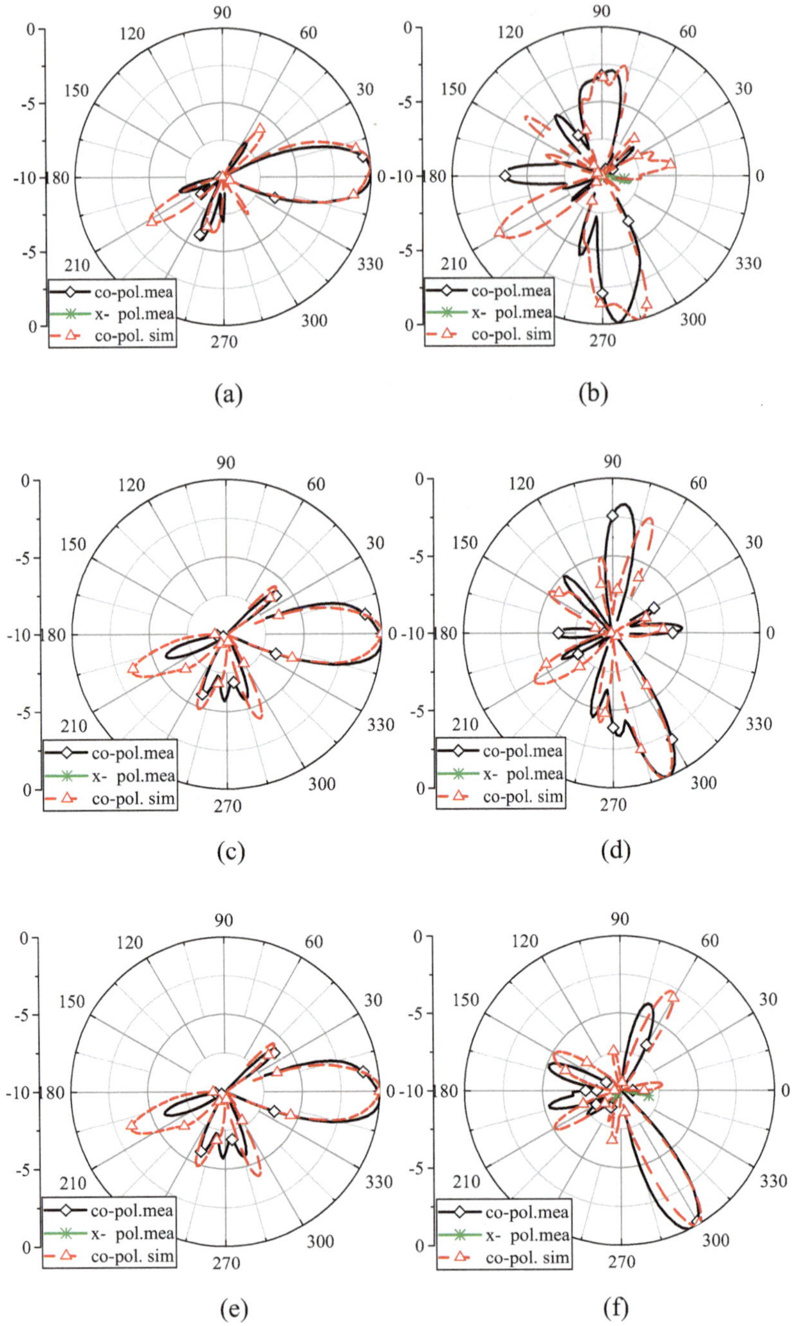

Fig. 8.22 Simulated and measured radiation patterns. **a** and **b** f = 5.3 GHz; **c** and **d** f = 5.5 GHz; **e** and **f** f = 5.7 GHz

Table 8.1 Performance of proposed antenna

Frequency (GHz)	Beam direction (degree)		Gain (dBi)	
	Endfire	Broadside	Endfire	Broadside
5.3	−2	84	6.7	3.0
5.5	2	68	7.2	3.6
5.7	4	62	6.5	2.7

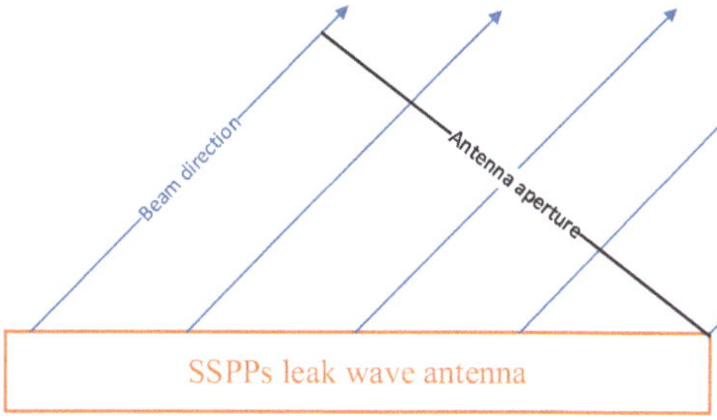

Fig. 8.23 The antenna aperture in broadside radiation

In addition, the suspended DC bias line will also affect the measurement results of the reconfigurable antenna. In addition, the material loss is also an important factor affecting the radiation efficiency and gain of the antenna.

Considering the above factors, the antenna is redesigned. The substrate plate adopts Arlon AD600, and the loss tangent is only 0.003. The DC bias line is printed on the back of the dielectric plate with the width of 0.15 mm to minimize the influence of DC bias line on AC. In addition, increasing the width of metal coplanar waveguide can reduce the back lobe of antenna. The transition structure is added in the transition part to improve the impedance matching and antenna efficiency.

The model diagram of the antenna after re-optimized design is shown in Fig. 8.25.

The far-field pattern of the re-optimized antenna at 5 GHz is shown in Fig. 8.26. The comparison of measured and simulated $|S_{11}|$ curves is shown in Fig. 8.27. The frequency band of $|S_{11}| < -10$ dB is 4.85–5.25 GHz, which can cover the WLAN frequency band of 5.2 GHz. The gain of the antenna is shown in Table 8.2. Compared with the gain of the reconfigurable SSPPs antenna in the previous section (Table 8.1), the gain of the re-optimized antenna is obviously improved, and the side radiation gain is increased by at least 2.5 dB. The dispersion curves of odd mode and even mode of the re-optimized SSPPs structure are shown in Fig. 8.28. It can be seen that the dispersion curves of the odd mode and the even mode coincide

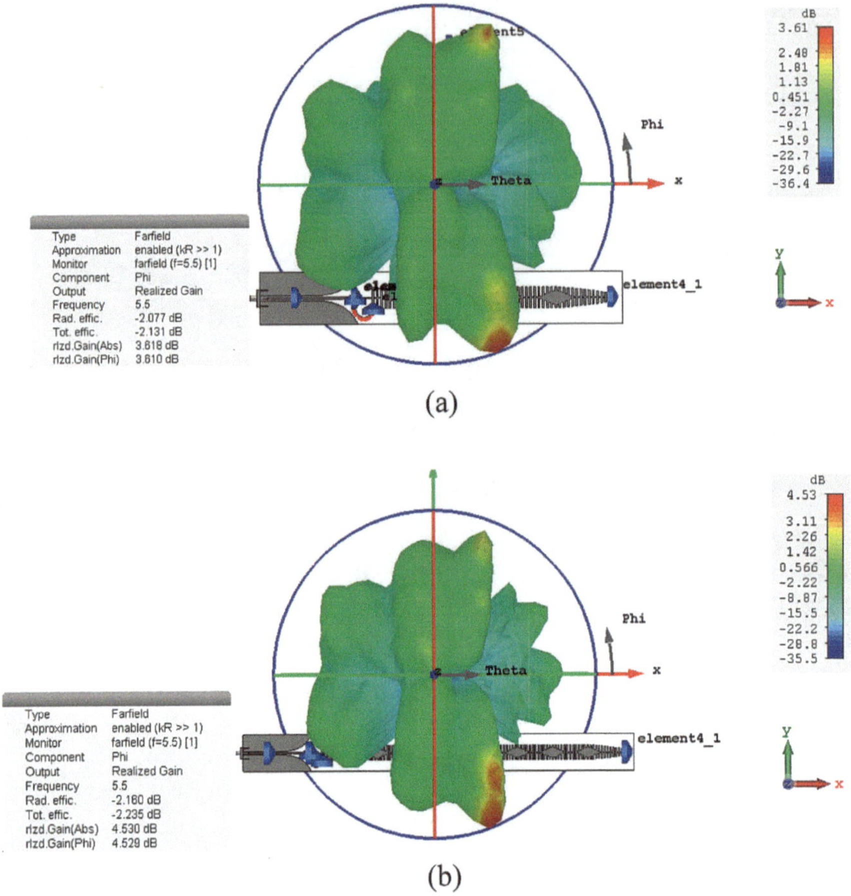

Fig. 8.24 Comparison of the radiation patterns at 5.5 GHz. **a** Antenna length is 218.7 mm; **b** antenna length is 341.7 mm

in a very narrow frequency band. This narrow band shows that the reconfigurable SSPPs antenna designed in this section can only work in this narrow band. The far-field pattern of the re-optimized antenna is shown in Fig. 8.29. The simulation results of the main lobe of the pattern are in good agreement with the measured results, but there are some deviations between the simulation and the measured patterns in the side radiation, which is mainly due to the errors in the measurement and processing. In addition, it can be seen that the re-optimized antenna has good pattern reconstruction ability in the working frequency band.

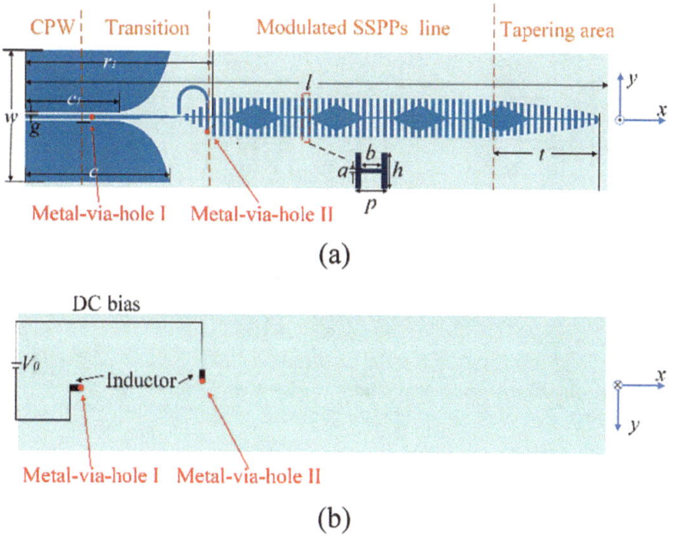

Fig. 8.25 Improved pattern reconfigurable antenna. **a** Front view, and **b** back view of the proposed antenna. Figure reproduced with permission from Ref. [5], © 2019 IEEE

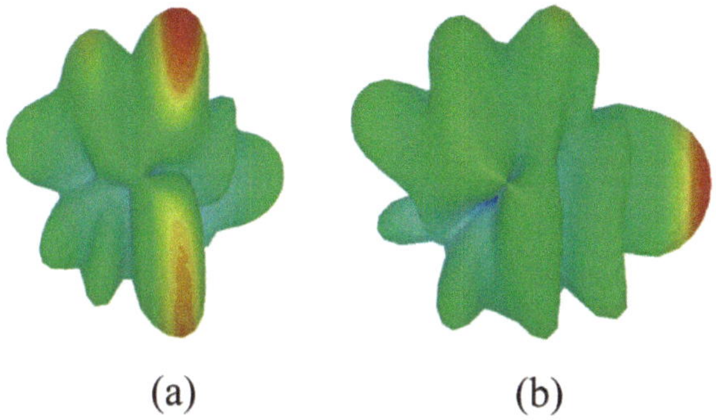

Fig. 8.26 Radiation pattern of improved pattern reconfigurable antenna at 5 GHz. **a** The broadside radiation pattern, and **b** the endfire radiation pattern. Figure reproduced with permission from Ref. [5], © 2019 IEEE

8.3 Conclusion

This chapter mainly studies a pattern reconfigurable antenna based on SSPPs structure. Through the RF switch constructed by PIN diode, the SSPPs antenna pattern can be quickly switched between side radiation and end fire modes. Therefore, it has good application value in the field of wireless communication and radar detection.

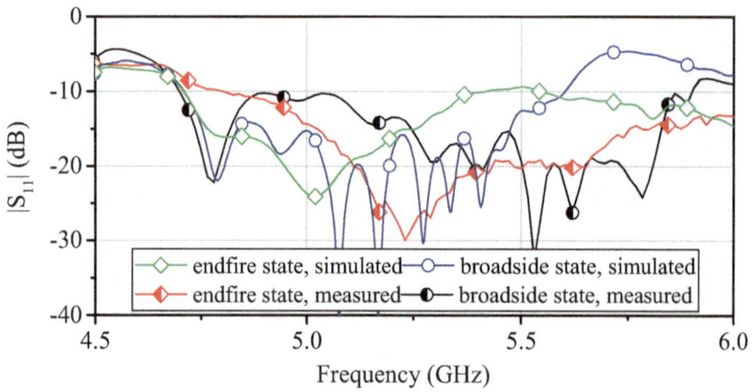

Fig. 8.27 Comparison of the simulated and measured $|S_{11}|$ values of the proposed pattern reconfigurable antenna. Figure reproduced with permission from Ref. [5], © 2019 IEEE

Table 8.2 Performance of improved antenna

Frequency (GHz)	Beam direction (degree)		Gain (dBi)	
	Endfire	Broadside	Endfire	Broadside
4.85	−4	+84, −92	9	6.6
5.05	0	+72, −80	8.5	6.1
5.25	+2	+56, −66	7.1	7.2

Table reproduced with permission from Ref. [5], © 2019 IEEE

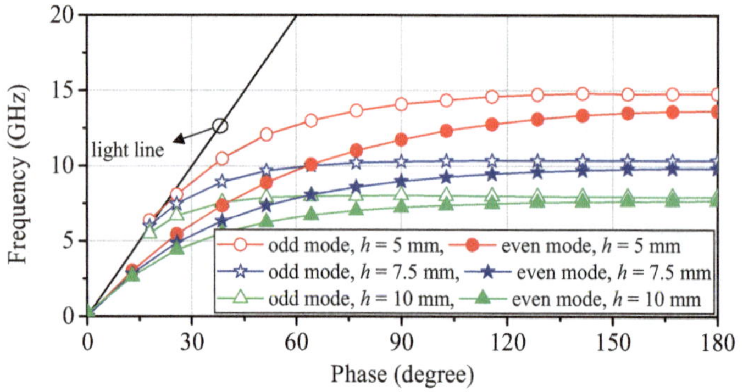

Fig. 8.28 Simulated dispersion relations of symmetric SSPPs structures with different groove depth for even mode and odd mode. Figure reproduced with permission from Ref. [5], © 2019 IEEE

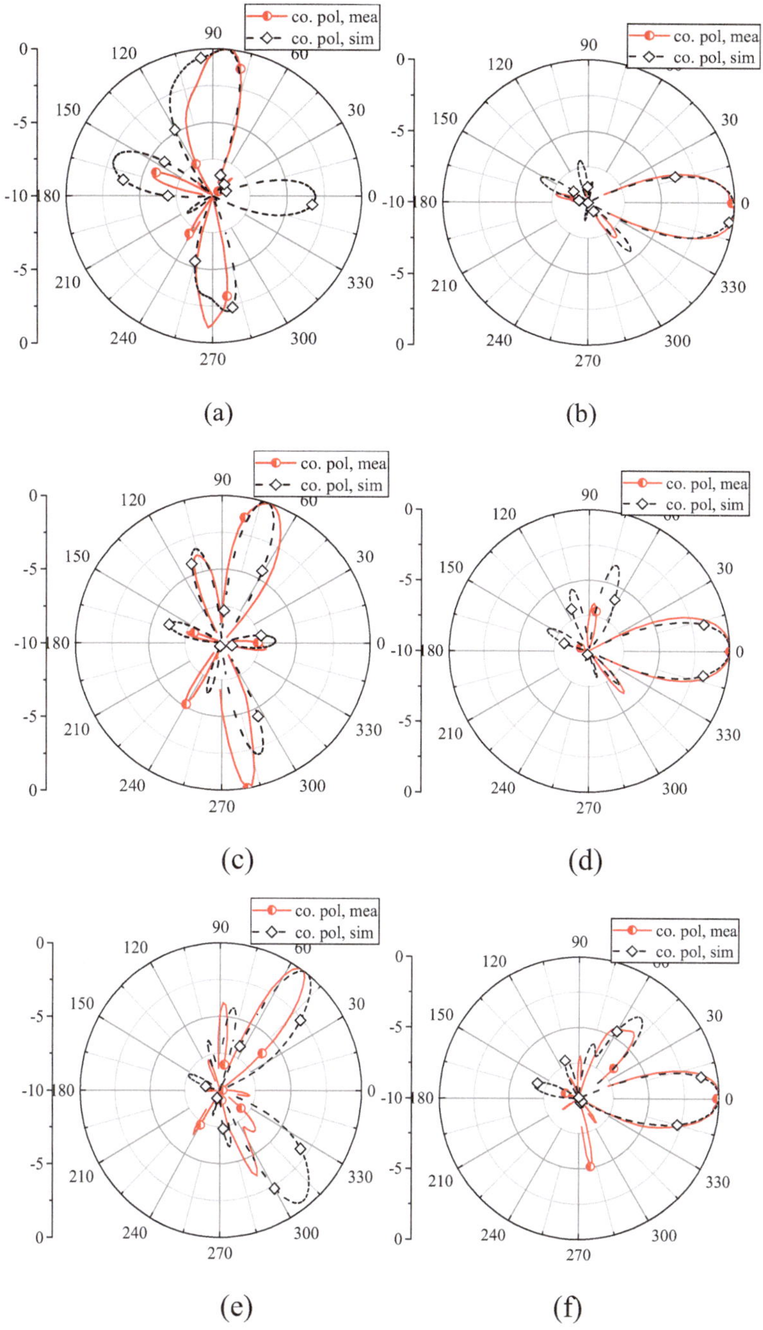

Fig. 8.29 Simulated and measured radiation patterns. **a** and **b** are at 4.85 GHz; **c** and **d** are at 5.05 GHz; **e** and **f** are at 5.25 GHz. Figure reproduced with permission from Ref. [5], © 2019 IEEE

References

1. Z. Liao et al., High-order localized spoof surface plasmon resonances and experimental verifications. Sci. Rep. **5**(1), 9590 (2015)
2. A. Kianinejad et al., Design and modeling of spoof surface plasmon modes-based microwave slow-wave trans-mission line. IEEE Trans. Microw. Theory Tech. **63**(6), 1817–1825 (2015)
3. A. Kandwal, Q. Zhang, X. Tang, L.W. Liu, G. Zhang, Low-profile spoof surface plasmon polaritons traveling-wave antenna for endfire radiation. IEEE Ant. Wirel. Propag. Lett. **17**(2), 184–187 (2017)
4. J.Y. Yin, D. Bao, J. Ren, H.C. Zhang, B.C. Pan, Y. Fan, T.J. Cui, Endfire radiations of spoof surface plasmon polaritons. IEEE Ant. Wirel. Propag. Lett. **16**, 597–600 (2017)
5. K. Zhuang et al., Pattern reconfigurable antenna applying spoof surface plasmon polaritons for wide angle beam steering. IEEE Access **7**, 15444–15451 (2019). https://doi.org/10.1109/access.2019.2895106
6. S.K. Gu, H.F. Ma, B.G. Cai, T.J. Cui, Continuous leaky-wave scanning using periodically modulated spoof plasmonic waveguide. Sci. Rep. **6**, 29600 (2016)
7. Q. Zhang, Q. Zhang, Y. Chen, Spoof surface plasmon polariton leaky-wave antennas using periodically loaded patches above PEC and AMC ground planes. IEEE Ant. Wirel. Propag. Lett. **16**, 3014–3017 (2017)
8. D. Liao, Y. Zhang, H.G. Wang, Wide-angle frequency-controlled beam scanning antenna fed by standing wave based on the cut-off characteristics of spoof surface plasmon polaritons. IEEE Ant. Wirel. Propag. Lett.
9. X. Gao, L. Zhou, T.J. Cui, Odd-mode surface plasmon polaritons supported by complementary plasmonic metamaterial. Sci. Rep. **5**(1) (2015). Art. No. 9250
10. X. Liu, Y. Feng, K. Chen, B. Zhu, J. Zhao, T. Jiang, Planar surface plasmonic waveguide devices based on symmetric corrugated thin film structures. Opt. Express **22**(17), 20107–20116 (2014)
11. G.S. Kong, H.F. Ma, B.G. Cai, T.J. Cui, Continuous leaky-wave scanning using periodically modulated spoof plasmonic waveguide. Sci. Rep. **6**(1) (2016). Art. No. 29600
12. H. Ya-Juan, Z. Jie-Qiu, L. Yong-Feng, W. Jia-Fu, Q. Shao-Bo, Z. An-Xue, 360° scanning multi-beam antenna based on spoof surface plasmon polaritons. Acta. Phys. Sin **65**(14), 147301 (2016)

Chapter 9
Phased-Mode Spoof Surface Plasmon Polaritons Antenna

Junping Geng, Jing Zhang, and Kun Wang

Harrington has proved that the maximum directivity coefficient D of a source region is a function of the number of multipole modes N ($D_{max} = N^2 + 2N$) [1, 2]. Theoretically, the larger the N is, the larger the directional coefficient D is, and the stronger the super directional needle radiation is. When the electric and magnetic dipoles are excited with equal amplitude and equal phase, the pattern is the heart shape in the electric and magnetic fields. The forward radiation is enhanced and the backward radiation is obviously suppressed. This structure is often called Huygens source. This kind of magnetoelectric orthogonal radiation element can obtain higher directional radiation gain. On the one hand, it preliminarily proves the correctness of Harritung's theory. On the other hand, it also shows that multiple modes can share the same aperture, and multiple modes can be superimposed with each other to form a comprehensive mode with shared aperture.

For the resonant structure, the mode is essentially the source distribution. In the most of the time, the mode is the aperture current standing wave distribution. When multiple modes share the same aperture, it is the vector superposition of the corresponding current standing wave distribution.

If two modes coexist on an aperture, the superposition of the two modes is the synthetic mode of the aperture. However, the superposition of the current distribution of the two modes is related to their respective initial phases. In other words, it is related to the phase difference between the two modes. Different phase

J. Geng (✉) · J. Zhang · K. Wang
Department of Electronic Engineering, Shanghai Jiao Tong University,
Shanghai 200240, China
e-mail: gengjunp@sjtu.edu.cn

J. Zhang
e-mail: sjtu_zhangjing@qq.com

K. Wang
e-mail: wanggkun@sjtu.edu.cn

J. Geng et al., *Spoof Surface Plasmon Polaritons Antenna*, Modern Antenna,
https://doi.org/10.1007/978-981-16-4721-5_9

differences will lead to different synthetic modes. We call it phased-mode aperture, just like phased array antenna.

Similarly, if this phased-mode aperture unit is a multi port antenna aperture, and each port excites one mode. When all ports are excited at the same time, the synthetic mode of aperture is the vector sum of the modes excited by each port, which is related to the phase difference between ports. Correspondingly, the radiation field of this aperture is also related to the phase difference between ports. So, this multi port antenna aperture unit is called phased-mode antenna.

Recently, spoof surface plasmonic polaritons (SSPPs, spoof surface plasmonic polaritons) is proposed, which can work in multipole mode. In theory, different excitation methods can stimulate different modes of odd mode and even mode in SSPPs structure. Our research shows that these two SSPPs modes can also be synthesized to form synthesised mode in aperture, and the characteristics of the synthesised mode are related to the difference between the original odd mode and even mode.

Based on the SSPPs antenna, this chapter proposes and studies the phased-mode SSPPs antenna, which provides a new idea and direction for the development of antenna technology and array antenna.

9.1 Broadside/End-Fire Pattern Reconfigurable Antenna Based Phased Mode Spoof Surface Plasmon Polaritons Antenna

Figure 9.1 is the geometry of the proposed phased-mode reconfigurable antenna based SSPPs structure [3]. It includes CPW feeding part, coupling part and SSPPs antenna. There are two SMA ports connect the single side CPW respectively. At the end of the CPW, the two strips coupled to SSPPs transmission line from two side. Then the couple EM field combined together, and transmits and radiates along the SSPPs radiation part that with gradually shallow gap. The SSPPs antenna is printed on the dielectric substrate (with relative dielectric constant of 3.66 and thickness of 0.762 mm).

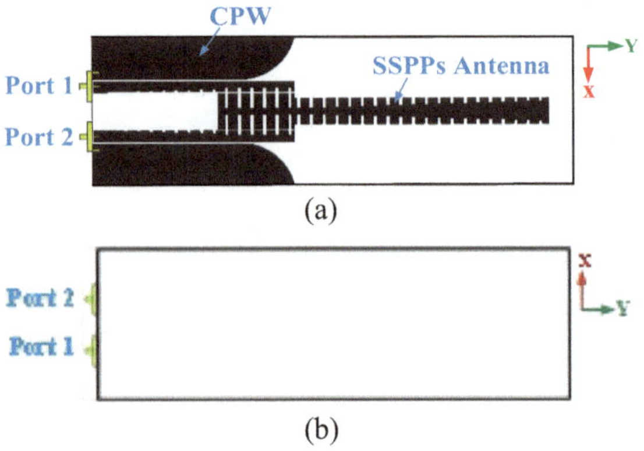

Fig. 9.1 Geometry of the proposed antenna. **a** Top view; **b** back view

It can achieve reconfigurable radiation patterns between broadside radiation and end-fire radiation. By controlling phase difference between two input excitation source, the surface current distribution on SSPPs structure will change simultaneously. The gradual shallow SSPPs transmission line can support odd mode with end-fire radiation and even mode with broadside radiation. When the phase difference between two ports is 0°, there will appear the symmetrical currents on the SSPPs antenna, which is even mode, and means broadside radiation pattern. When the phase difference between two ports is 180°, there will appear the anti-symmetrical currents on the SSPPs antenna, which is odd mode, and means end-fire radiation pattern.

Figure 9.2 shows the simulated return loss and port isolation curves. The impedance bandwidth with $|S_{11}| \leftarrow 10$ dB and $|S_{12}| \leftarrow 10$ dB is from 7.37 to 8.33 GHz, and the relative bandwidth is about 12.23%. The isolation between the Port 1 and Port 2 is very larger, which is higher than 24.9 dB in the operation frequency band.

In Fig. 9.3, The radiation patterns of the proposed antenna with endfire in the odd mode at 7.4, 7.7, 8.0 and 8.3 GHz are depicted. The endfire radiation patterns is in good symmetry, and the 3-dB beam width is about 35°. But the side lobe increase with the frequency, and it is even larger than the center lobe at 8.3 GHz.

The radiation patterns of the proposed SSPPs antenna with broadside radiation in the even mode at 7.4, 7.7, 8.0 and 8.3 GHz are shown in Fig. 9.4. The patterns are symmetrical and there are obvious null in the symmetrical axis direction of the antenna.

It is obvious, the broadside radiation and end-fire radiation pattern reconfigurable antenna based on SSPPs structure is realized by simulation. Though controlling the phase difference between two input excitation source, the reconfigurable radiation patterns are obtained. The proposed antenna can work in endfire (in the odd mode) and broadside (in the even mode), respectively. The proposed antenna has a 12.23% impedance bandwidth with covering 7.37 to 8.33 GHz and the isolation of 24.9 dB. The proposed pattern reconfigurable SSPPs antenna work in two states phased mode, which maybe very useful in wireless communication system.

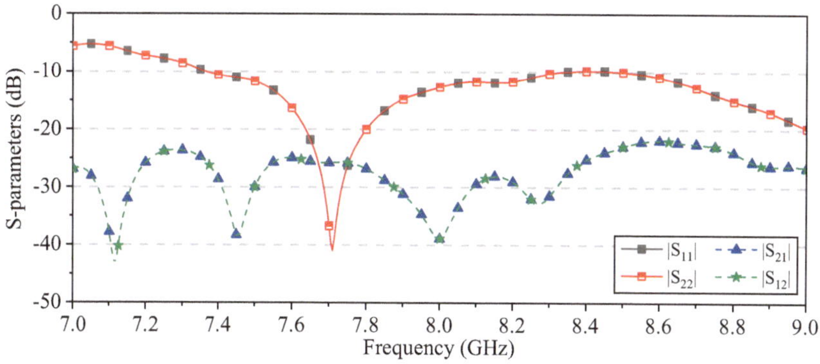

Fig. 9.2 Reflection coefficient and port isolation

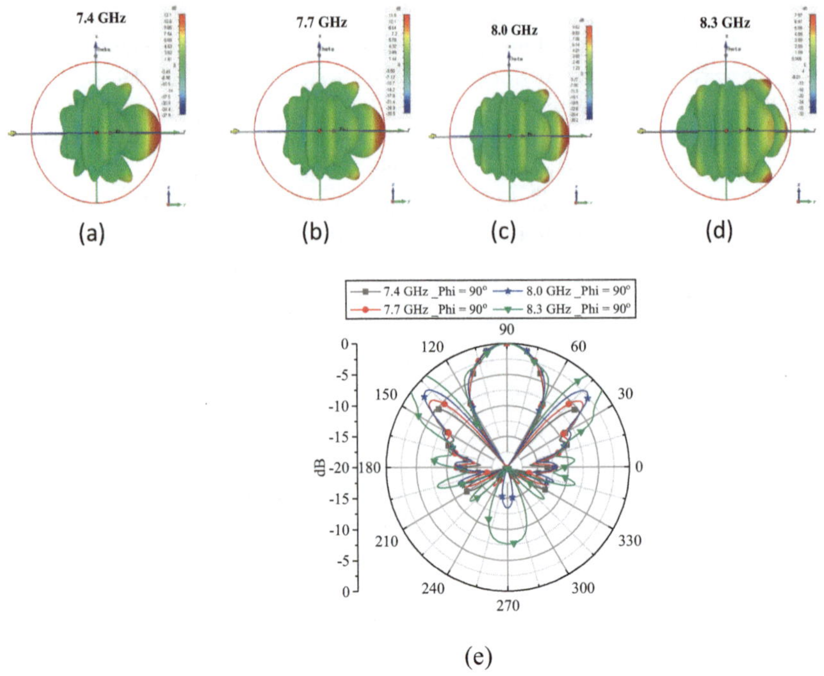

Fig. 9.3 Endfire radiation patterns of SSPPs antenna

9.2 Phased-Mode Spoof Surface Plasmon Polaritons Antenna

As an extension of surface plasmon polaritons (SPPs) in the optical frequencies, SSPPs structure has been widely applied to broadband absorbers [4, 5], microwave circuits [6], antennas [7–11], polarization modulators [12] and dynamically tunable integrated devices [13, 14]. Also, SSPPs structure has great application potential in future beam scanning technologies due to their inherent properties of strong field confinement and energy localization. Generally, there are two fundamental modes supported on SSPPs structure, even mode and odd mode. End-fire beams with high gain are provided by traveling waves of odd mode, while broadside beams are radiated by standing waves of even mode, which makes it an ideal candidate for wide-angle scanning.

Usually, only one mode can be excited by the single-port SSPPs antenna. In [15], the SSPPs antenna is designed for radar cross section (RCS) reduction performances by placing a monopole at a certain distance beside the SSPPs structure for coupling and feeding, where merely odd mode is excited. Then the antenna is extended to a 5 $\times$ 5 array, which performs high gain and low RCS characteristics, but no scanning. A pattern reconfigurable antenna applying SSPPs is proposed in

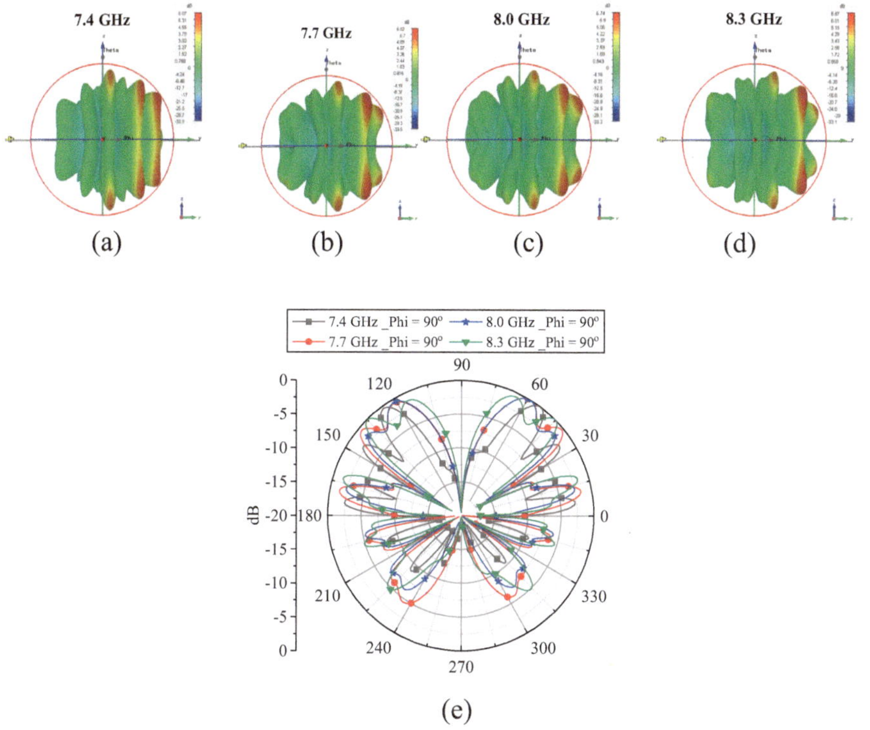

Fig. 9.4 Broadside radiation patterns of SSPPs antenna

[16], which innovatively realizes the radiation switching between odd mode and even mode of SSPPs structure under single port condition. It supports the end-fire lobe at odd mode and the quasi-omnidirectional lobe at even mode, but lacks the intermediate states between those two modes, and is not competent for beam scanning.

In this section, a phase-mode SSPPs antenna with dual-port is proposed [17], which can work in odd mode, even mode and mixed modes with the characteristic of spatial lobe scanning simply by varying the phase difference between the two ports of one antenna element. Hence, the presented antenna is qualified to cover 1/8 spherical space by lobe scanning in the band of 4.6–5.6 GHz.

9.2.1 Antenna Configuration

The presented phased-mode SSPPs antenna with lobe scanning dual-port feeding is given in Fig. 9.5. It is printed on an Arlon AD255 substrate with the thickness of 1.524 mm (ε_r = 2.55 and tanδ = 0.0018). The antenna includes two trapezoidal

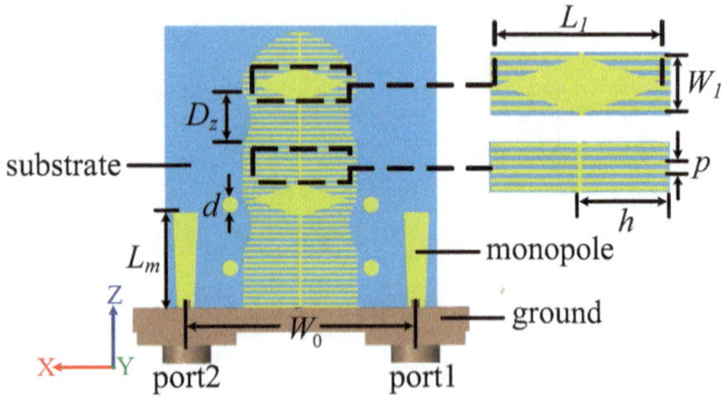

Fig. 9.5 Configuration of the SSPPs antenna with phased-mode. Figure reproduced with permission from Ref. [17], © 2021 IEEE

monopoles, a double side SSPPs structure modulated by two rhombs in center, and four small circular patches load on both side of the SSPPs structure, and a 3-mm-thick ground made of aluminum. The trapezoidal monopoles are fed by two 50 Ω SMA connector respectively. Then, the electromagnetic (EM) wave is coupled into the SSPPs structure from the trapezoidal monopole. The EM waves propagate along the symmetric SSPPs structure and radiate outward from the SSPPs structure too.

The ground is placed perpendicular to the SSPPs plane, of which size is 43 mm × 30 mm × 3 mm. The size of the Arlon AD255 substrate is 35.6 mm 40 mm × 1.524 mm as shown in Fig. 9.5, the dimensions of the rhombic patches are L_1 and W_1. The length of the trapezoidal monopole is L_m, which is about $\lambda/4$ (λ is the equivalent wavelength with substrate at central operating frequency 5.2 GHz). The period and width of the arm of the SSPPs structure are p and a. There are two pairs of parabolic grooves are notched on both sides of the SSPPs structure, which is good to the gain and the end-fire radiation.

CST Microwave Studio software is used to simulate the antenna. The initial parameters of the antenna are $D_z = 5$ mm, $d = 3$ mm, $L_m = 13$ mm, $L_1 = 10.5$ mm, $a = 0.4$ mm, $p = 0.8$ mm, $h = 8$ mm, $W_1 = 4$ mm.

9.2.2 *Working Principle for Phased-Mode Spoof Surface Plasmon Polaritons Antenna*

Figure 9.6 is the schematic diagram of the SSPPs antenna with phased-mode, dual ports are fed with the same amplitude but different phase $\Delta\varphi$, φ_1 and φ_2 are the incident signal phases of port1 and port2 respectively, and the phase difference between the two ports is $\Delta\varphi = \varphi_2 - \varphi_1$.

In SSPPs waveguide, odd mode and even mode are two fundamental transmission modes in SSPPs structure, which mainly depends on the surface current distribution on SSPPs structure. Here, as the phase difference $\Delta\varphi$ between the two ports changes, the surface current distribution on the SSPPs structure changes.

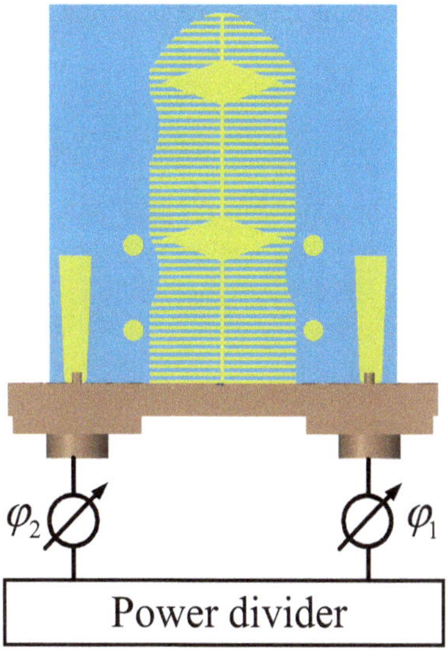

Fig. 9.6 Schematic diagram of the SSPPs antenna with phased-mode. Figure reproduced with permission from Ref. [17], © 2021 IEEE

Figure 9.7 shows the diagram of the current distributions on the two arms of SSPPs structure in two modes. In the odd mode, the current flow on the left armand right arm are same, as shown in Fig. 9.7a. However, since the monopole is located on the one side of the SSPPs structure, the magnitudes of current distribution on the left and right arm are different when single port is excited. Here, define $J_{1l}(x,z)$ and $J_{1r}(x,z)$ as the magnitudes of the current on the left and right arms when only port1 is excited, while $J_{2l}(x,z)$ and $J_{2r}(x,z)$ are only port2 being excited. $\xi_1(x,z)$ and $\xi_2(x,z)$ are defined as the spatial phases from port1 or port2 to the SSPPs arms, respectively. The two ports are assumed to be isolated enough, then the current distribution of SSPPs structure when only port1 is excited can be expressed as

$$\vec{J_1} = \vec{a_x} \cdot [J_{1l}(x,z) + J_{1r}(x,z)] \cdot e^{j\xi_1(x,z)} \tag{9.1}$$

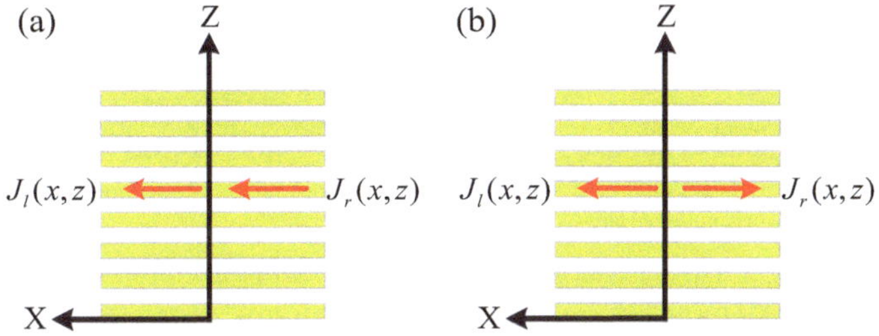

Fig. 9.7 Current distributions in the SSPPs structure **a** odd mode; **b** even mode. Figure reproduced with permission from Ref. [17], © 2021 IEEE

And the current distribution of the case that only port2 is excited can be written as

$$\vec{J_2} = \vec{a_x} \cdot [J_{2l}(x,z) + J_{2r}(x,z)] \cdot e^{j\xi_2(x,z)} \cdot e^{j\pi} \tag{9.2}$$

When port1 and port2 are fed with opposite phase separately, the current amplitude distributions are antisymmetric.

$$J_{1l}(x,z) = J_{2r}(x,z) = J_a(x,z) \tag{9.3}$$

$$J_{1r}(x,z) = J_{2l}(x,z) = J_b(x,z) \tag{9.4}$$

When two ports are ecited simutaneouslywith initial phases φ_1 and φ_2 respectively, the total current distribution on the SSPPs arm is:

$$\begin{aligned} \vec{J_t} &= \vec{J_1}e^{j\varphi_1} + \vec{J_2}e^{j\varphi_2} \\ &= \vec{a_x} \cdot [J_{1l}(x,z) + J_{1r}(x,z)] \cdot e^{j\xi_1(x,z)} \cdot e^{j\varphi_1} \\ &\quad + \vec{a_x} \cdot [J_{2l}(x,z) + J_{2r}(x,z)] \cdot e^{j\xi_2(x,z)} \cdot e^{j\pi} \cdot e^{j\varphi_2} \\ &= \vec{a_x} \cdot [J_a(x,z) + J_b(x,z)] \cdot e^{j\varphi_1} \cdot \left[e^{j\xi_1(x,z)} - e^{j\xi_2(x,z)}e^{j\Delta\varphi}\right] \end{aligned} \tag{9.5}$$

When $\varphi_1 = \varphi_2$, $\Delta\varphi = 0°$, it means same phase feeding, and the total current distribution can be achieved:

$$\vec{J}_{to} = \vec{a_x} \cdot [J_a(x,z) + J_b(x,z)] \cdot e^{j\varphi_1} \cdot \left[e^{j\xi_1(x,z)} - e^{j\xi_2(x,z)}\right] \tag{9.6}$$

Equation (6) means the current directions on the left and right arms are opposite at $\Delta\varphi = 0°$, which can be conclude that the antenna works in even mode.

When two ports are opposite phase fed, $\Delta\varphi = 180°$, the total current distribution on the SSPPs structure is:

$$\vec{J}_{t180} = \vec{a_x} \cdot [J_a(x,z) + J_b(x,z)] \cdot e^{j\varphi_1} \cdot \left[e^{j\xi_1(x,z)} + e^{j\xi_2(x,z)}\right] \tag{9.7}$$

It is obvious, the current directions on the left and right arms are the same at $\Delta\varphi = 180°$, it means that the antenna works in odd mode.

Simulated E_x component distribution varies with the phase difference $\Delta\varphi$ between the two ports is depicted in Fig. 9.8. When only one port works, the SSPPs structure acts as a director, and the traveling waves is dominant odd mode being excited, as shown in Fig. 9.8f, g, generating end-fire lobe. If two ports work at the same time with different $\Delta\varphi$, odd mode, even mode and mixed modes of the two are generated. When $\Delta\varphi = 0°$, the symmetric magnitudes of electric field and reversed directions are shown in Fig. 9.8a, which isa typical even mode SSPPs characteristic. When $\Delta\varphi = 180°$, both electric field magnitudes and phases are the same on

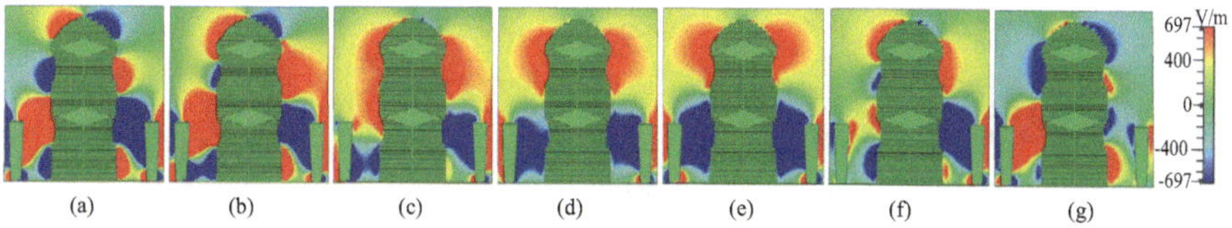

Fig. 9.8 Simulated E_x component varies with the phase difference between the two ports of the proposed antenna. **a** $\Delta\varphi = 0°$; **b** $\Delta\varphi = 60°$; **c** $\Delta\varphi = 90°$; **d** $\Delta\varphi = 135°$; **e** $\Delta\varphi = 180°$; **f** only port1 excited; **g** only port2 excited. Figure reproduced with permission from Ref. [17], © 2021 IEEE

the left and right arms in Fig. 9.8e, which means that the antenna generates odd SSPPs waves. As shown in Fig. 9.8b–d, corresponding to $\Delta\varphi = 60°$, 90° and 135°, the electric field distributions are in transitional states.

Corresponding to different phase difference $\Delta\varphi$ in Fig. 9.8, the simulated power flow viewed from the top are shown in Fig. 9.9 under different phase differences $\Delta\varphi$. Figure 9.9b shows the power flow mainly pointing to the +*y* and −*y* directions at $\Delta\varphi = 0°$, which corresponds to the even mode. Then, at $\Delta\varphi = 60°$, 90° and 135°, the power flow distributions are in transitional states as depicted in Fig. 9.9c–e. Ultimately, At $\Delta\varphi = 180°$, Fig. 9.9f shows the power flow pointing to the +*z* direction, which is clear in the odd mode. It is noted that with the phase difference gradually increasing from 0° to 180°, the power flow main trend direction rotates from 45° to the +*x* direction in the horizontal direction, and then to the end of SSPPs structure in the vertical direction. When single one port is fed,the power flow of the proposed antenna are given in Fig. 9.9g, h, in which the energy is efficiently coupled to SSPPs structure, but less to another side monopole, and verifies the isolation between two ports.

3D patterns at different phase difference $\Delta\varphi$ and single port excitation are illustrated in Fig. 9.10. Obviously, as the phase difference $\Delta\varphi$ varies, the radiation patterns can scan in 1/8 of the spherical space where $X > 0$, $Y > 0$, and $Z > 0$.

The 3 dB beam width of main lobe the (Φ, θ) coordinate spaceare shown in Fig. 9.11. Unlike traditional scanning antennas, the proposed SSPPs antenna is able to scan large angle in (Φ, θ) space with the phase difference, and the 3 dB beam widths are competent to cover a wide range. So, it is called phased-mode SSPPs antenna.

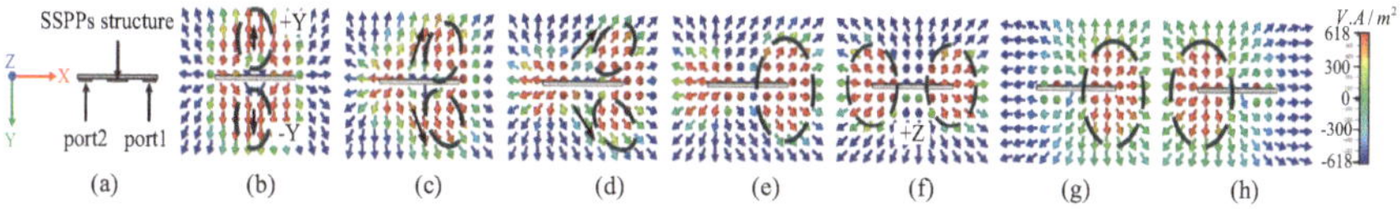

Fig. 9.9 Simulated power flow varies with the phase difference between the two ports of the proposed antenna. **a** Top view of the proposed antenna; **b** $\Delta\varphi = 0°$; **c** $\Delta\varphi = 60°$; **d** $\Delta\varphi = 90°$; **e** $\Delta\varphi = 135°$; **f** $\Delta\varphi = 180°$; **g** only port1 excited; **h** only port2 excited. Figure reproduced with permission from Ref. [17], © 2021 IEEE

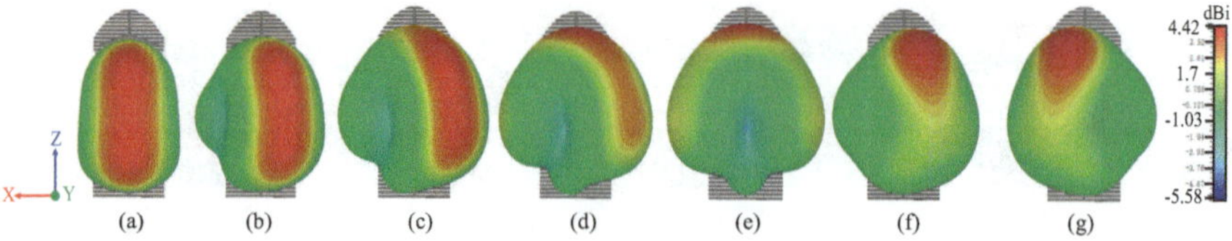

Fig. 9.10 Simulated 3D pattern varies with the phase difference between the two ports of the proposed antenna. **a** $\Delta\varphi = 0°$; **b** $\Delta\varphi = 60°$; **c** $\Delta\varphi = 90°$; **d** $\Delta\varphi = 135°$; **e** $\Delta\varphi = 180°$; **f** only port1 excited; **g** only port2 excited. Figure reproduced with permission from Ref. [17], © 2021 IEEE

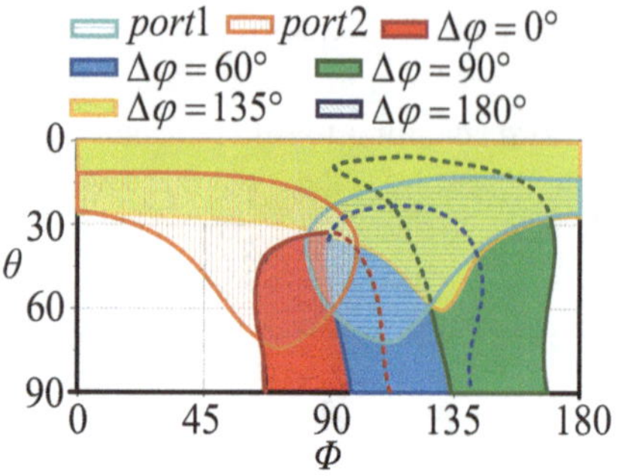

Fig. 9.11 Scanning diagram in (Φ, θ) coordinate space corresponding to Fig. 9.10. Figure reproduced with permission from Ref. [17], © 2021 IEEE

9.2.3 *Comparison Between Measurement and Simulation Results*

After simulation and optimization, the optimal parameters of the antenna are listed in Table 9.1. As shown in Fig. 9.12, the prototype of the proposed s antenna was manufactured accordding to the optimized size, and it was tested in SATIMO SG64 3D microwave anechoic chamber. As shown in Fig. 9.6, the power splitter and phase shifters are introduced to construct the phased-mode antenna. And the phase shifters are used for phase difference control.

The measured $|S_{11}|$ and $|S_{21}|$ curves are given in Fig. 9.13. The measured impedance bandwidth covers 4.5–6 GHz, about 28.6%. It is obvious that the measured return loss $|S_{11}|$ are consistent with the simulated results in trend, but larger than the simulated one, which is attributed to manufacture errors.

Table 9.1 Optimized parameter values

Parameters	D_z	d	L_m	L_1
Value (mm)	6.2	2	14.4	11.8
Parameters	a	P	H	W_1
Value (mm)	0.34	0.7	7	39

Table reproduced with permission from Ref. [17], © 2021 IEEE

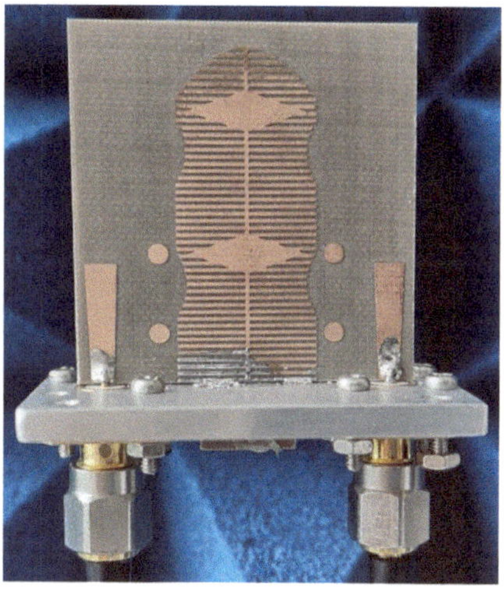

Fig. 9.12 Fabricated prototype of the proposed antenna. Figure reproduced with permission from Ref. [17], © 2021 IEEE

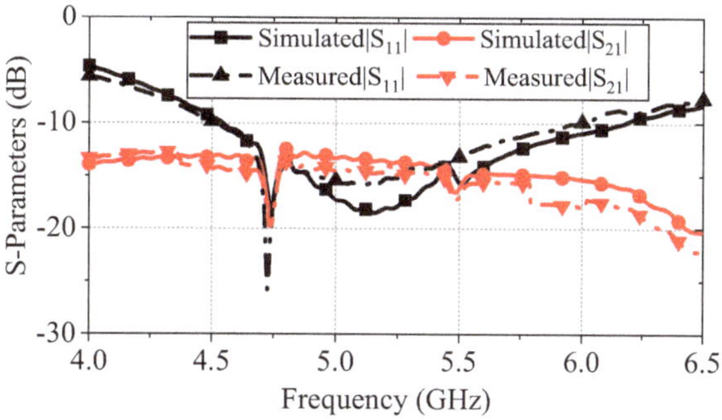

Fig. 9.13 The simulated and measured S-parameters. Figure reproduced with permission from Ref. [17], © 2021 IEEE

The measured gains with different phase difference at 5.2 GHz are depicted in Fig. 9.14. At 5.2 GHz, for the phase difference $\Delta\varphi$ varies from 0° to 180°, the measured gain is 3.4–4.4 dBi, which is basically in agreement with the simulated.

In order to show the spatial scanning performances of the proposed antenna, the measured spatial patterns projected in the (Φ, θ) space are illustrated in Fig. 9.15,

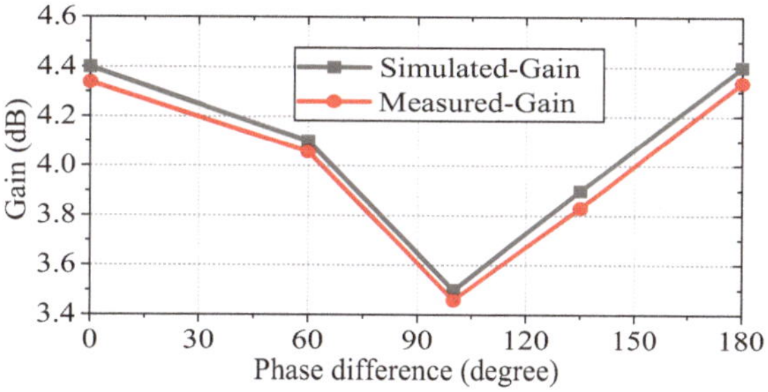

Fig. 9.14 The simulated and measured gains varying with phase difference at 5.2 GHz. Figure reproduced with permission from Ref. [17], © 2021 IEEE

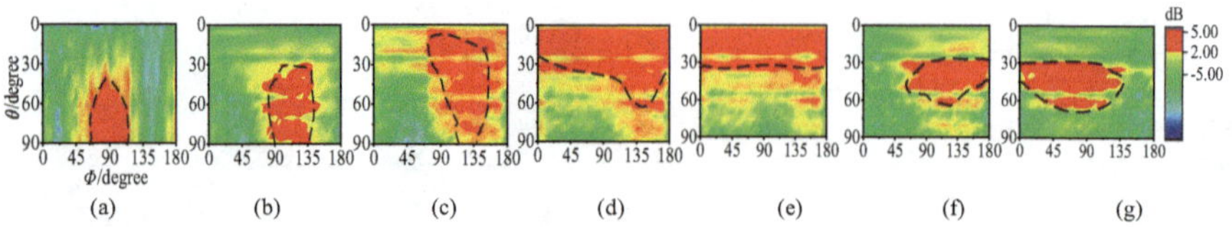

Fig. 9.15 Measured patterns in (Φ, θ) coordinate space varies with the phase difference between the two ports of the proposed antenna. **a** $\Delta\varphi = 0°$; **b** $\Delta\varphi = 60°$; **c** $\Delta\varphi = 90°$; **d** $\Delta\varphi = 135°$; **e** $\Delta\varphi = 180°$; **f** only port1 excited; **g** only port2 excited. Figure reproduced with permission from Ref. [17], © 2021 IEEE

respectively. And the 3 dB beam width coverage is represented by the red area in Fig. 9.15. The simulated patterns with different $\Delta\varphi$ are shown in Fig. 9.16. At $\Delta\varphi = 0°$, the measured main lobe in Fig. 9.15a corresponds to the simulated even mode of the the 3D pattern of Fig. 9.10a and its project in Fig. 9.16a. It is clear that the main lobe is uniformly scanned from $\Phi = 90°$ to $\Phi = 120°$ along the Φ-direction as $\Delta\varphi$ varies from 0° to 90°. Then, when $\Delta\varphi$ increase from 135° to 180°, the main lobe shows clear scanning trend along θ-direction. At $\Delta\varphi = 180°$, the radiation pattern shows end-fire beam towards +z axis in odd mode. It is noted that varying the phase difference $\Delta\varphi$ results in the main lobe of the antenna scanning gradually from the broadside radiation pattern in the even mode to transition pattern in the mixed modes, and then to the end-fire lobe in the odd mode, which is a continuous varying process in the 1/8 spherical space of $X > 0$, $Y > 0$, and $Z > 0$. So, it is called phased-mode antenna.

The comparison between the proposed phased-mode SSPPs antenna and other reported SSPPs antennas and phased-mode antenna are listed in Table 9.2. The proposed antenna can realize continuous transition from broadside lobe to end-fire lobe by continuesly modes varying, which supports the 3D scanning ability with more diverse.

Compared with traditional antennas, the proposed phased-mode antenna has the following two outstanding features:

1. The dual-port feeding mode and phase difference excitation make the antenna element itself have multi-dimensional beam scanning characteristics. It has a wide range of application potential in the phased array.

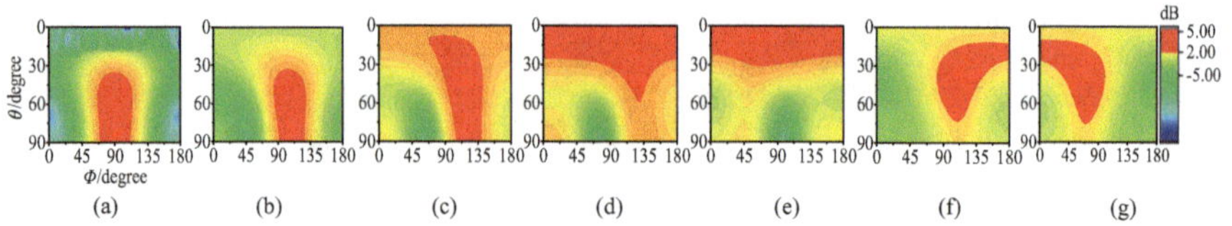

Fig. 9.16 Simulated patterns in (Φ, θ) coordinate space varies with the phase difference between the two ports of the proposed antenna. **a** $\Delta\varphi = 0°$; **b** $\Delta\varphi = 60°$; **c** $\Delta\varphi = 90°$; **d** $\Delta\varphi = 135°$; **e** $\Delta\varphi = 180°$; **f** only port1 excited; **g** only port2 excited. Figure reproduced with permission from Ref. [17], © 2021 IEEE

Table 9.2 Comparison with other reported SSPPs antenna and phase-mode antenna

Ref.	Working mode	Number of modes	Scanning principle	Main lobe scanning range	3 dB coverage scanning range
[16]	Even mode and odd mode	2	Cannot scan	–	–
[18]	Even mode	1	Frequency scanning	1D, −14.3° to +18.6°	–
[19]	Phased mode	Continuous multiple modes	Phased mode scanning	1D, −60° to +60°	–
[20]	Odd mode	1	Frequency scanning	1D, −60° to +63°	–
Proposed	Phaaed-mode: even mode, odd mode and mixed mode	Continuous multiple modes	Phased mode scanning	3D, Φ: +90° ~ +120° θ: 0° ~ +60°	3D, 0° < Φ < +180°, 0° < θ < +90°, except 0° < Φ < +70° and 30° < θ < + 90°

2. The continuous conversion and large-angle switching from the broadside lobe to the end-fire lobe are realized. The intermediate state is not limited, so the beam coverage is wide.

9.3 Summary of This Chapter

The phased-mode antennas based on the dual-port SSPPs antenna are proposed in this chapter. The designed phased-mode SSPPs antennas don't only realize the broadside radiation pattern in even mode and end fire pattern in odd mode, but also can realize the continusely scanning transition radiation patterns in the mixed modes with continusly varying the phase difference between two ports. The phased-mode antenna provides a new method and direction for the development of antenna technology.

References

1. R.F. Harrington, On the gain and beamwidth of directional antennas. IRE Trans. Antennas Propag. **6**, 219 (1958)
2. R.F. Harrington, Effect of antenna size on gain, bandwidth, and efficiency. J. Res. Natl. Bur. Stand. **64D**, 1 (1960)

3. K. Wang et al., in *Pattern Reconfigurable Antenna Applying Spoof Surface Plasmon Polaritons*. 2019 13th European Conference on Antennas and Propagation (EuCAP), Krakow, Poland, 2019, pp. 1–2
4. Y. Pang et al., Spatial k-dispersion engineering of spoof surface plasmon polaritons for customized absorption. Sci. Rep. **6** (Art. no. 29429) (2016)
5. R.K. Jaiswal, N. Pandit, N.P. Pathak, Amplification of propagating spoof surface plasmon polaritons in ring resonator-based filtering structure. IEEE Trans. Plasma Sci. **48**(9), 3253–3260 (2020)
6. X. Shena, T.J. Cui, D. Martin-Canob, J.F. Garcia-Vidal, Conformal surface plasmons propagating on ultrathin and flexible films. Proc. Nat. Acad. Sci. USA **110**(1), 40–45 (2013)
7. X. Zhang, J. Fan, J. Chen, High gain and high-efficiency millimeter-wave antenna based on spoof surface plasmon polaritons. IEEE Trans. Antennas Prop. **67**(1), 687–691 (2019)
8. Y. Fan et al., Frequency scanning radiation by decoupling spoof surface plasmon polaritons via phase gradient metasurface. IEEE Trans. Antennas Propag. **66**(1), 203–208 (2018)
9. G.S. Kong, H.F. Ma, B.G. Cai, T.J. Cui, Continuous leaky-wave scanning using periodically modulated spoof plasmonic waveguide. Sci. Rep. **6** (Art. no. 29600) (2016)
10. S. Li, Q. Zhang, Z. Xu, H. Zhao, X. Yin, Phase transforming based on asymmetric spoof surface plasmon polariton for endfire antenna with sum and difference beams. IEEE Trans. Antennas Propag. **68**(9), 6602–6613 (2020)
11. C. Wang et al., Frequency-selective structure with transmission and scattering deflection based on spoof surface plasmon polariton modes. IEEE Trans. Antennas Propag. **67**(10), 6508–6514 (2019)
12. Y. Li et al., Wideband polarization conversion with the synergy of waveguide and spoof surface plasmon polariton modes. Phys. Rev. Appl. **10**(6) (Art. no. 064002) (2018)
13. Z. Chen, W. Lu, Z. Liu, A. Zhang, B. Wu, H. Chen, Dynamically tunable integrated device for attenuation, amplification, and transmission of SSPP using grapheme. IEEE Trans. Antennas Propag. **68**(5), 3953–3962 (2020)
14. A. Zhang, W. Lu, Z. Liu, B. Wu, H. Chen, Flexible and dynamically tunable attenuator based on spoof surface plasmon polaritons waveguide loaded with grapheme. IEEE Trans. Antennas Propag. **67**(8), 5582–5589 (2019)
15. Y. Han, J. Wang, S. Gong, Y. Li, Y. Zhang, J. Zhang, Low RCS antennas based on dispersion engineering of spoof surface plasmon polaritons. IEEE Trans. Antennas Propag. **66**(12), 7111–7116 (2018)
16. K. Zhuang et al., Pattern reconfigurable antenna applying spoof surface plasmon polaritons for wide angle beam steering. IEEE Access **7**, 15444–15451 (2019)
17. J. Zhang, J. Geng, H. Zhou, C. Ren, K. Wang, S. Yang, W. Gao, E. Liu, X. Cheng, C. He, X. Liang, R. Jin, Dual-port phase antenna and its application in 1-D arrays to 2-D scanning. IEEE Trans. Antennas Propag. (2021). https://doi.org/10.1109/TAP.2021.3076558
18. H. Chen et al., Wideband frequency scanning spoof surface plasmon polariton planar antenna based on transmissive phase gradient metasurface. IEEE Antennas Wirel. Propag. Lett. **17**(3), 463–467 (2018)
19. E. Liu, J. Geng et al., Generalized principle of pattern multiplication based on the phase antenna element. Proc. IEEE AP-S Soc. Int. Symp. 353–354 (2020)
20. S. Xu et al., A Wide-angle narrowband leaky-wave antenna based on substrate integrated waveguide-spoof surface plasmon polariton structure. IEEE Antennas Wirel. Propag. Lett. **18** (7), 1386–1389 (2019)